Lecture Notes in Geoinformation and Cartography

The Lecture Notes in Geoinformation and Cartography series provides a contemporary view of current research and development in Geoinformation and Cartography, including GIS and Geographic Information Science. Publications with associated electronic media examine areas of development and current technology. Editors from multiple continents, in association with national and international organizations and societies bring together the most comprehensive forum for Geoinformation and Cartography.

The scope of Lecture Notes in Geoinformation and Cartography spans the range of interdisciplinary topics in a variety of research and application fields. The type of material published traditionally includes:

- proceedings that are peer-reviewed and published in association with a conference;
- post-proceedings consisting of thoroughly revised final papers; and
- research monographs that may be based on individual research projects.

The Lecture Notes in Geoinformation and Cartography series also includes various other publications, including:

- tutorials or collections of lectures for advanced courses;
- contemporary surveys that offer an objective summary of a current topic of interest; and
- emerging areas of research directed at a broad community of practitioners.

Noamen Rebai · Aniss Moumen ·
Mohamed El Bouhaddioui
Editors

GIS, Applied Computing and Data Science for Water Management

Selected Papers of the 4th International Conference GIRE3D Participatory and Integrated Management of Water Resources in Arid Zones

Springer

Editors
Noamen Rebai
Department of Geomatics
National Engineering School of Tunis
Tunis, Tunisia

Mohamed El Bouhaddioui
Department of Geology
National School of Mines Rabat
Rabat, Morocco

Aniss Moumen
IT, Logistics and Mathematics
National School of Applied Sciences
in Kénitra
Kénitra, Morocco

ISSN 1863-2246 ISSN 1863-2351 (electronic)
Lecture Notes in Geoinformation and Cartography
ISBN 978-3-031-63037-8 ISBN 978-3-031-63038-5 (eBook)
https://doi.org/10.1007/978-3-031-63038-5

This Springer imprint is published by the registered company Springer Nature Switzerland AG
The registered company address is: Gewerbestrasse 11, 6330 Cham, Switzerland

Preface

Water quality and scarcity reflect the mismanagement of this resource, resulting in increasing calls for protection and conservation of water resources and reconsidering the traditionally used management and planning tools for water resources. New technical and non-technical tools have been developed to efficiently manage water resources and ensure sustainable use and consumption of these resources. These tools include water quality assessment and monitoring coupled with methods for modeling, optimization, simulation, geographic information system (GIS), remote sensing, data mining, big data analysis, conflict resolution, and agent-based models.

A GIS is an information system dealing with geographically referenced data. It is a computer-based information system that allows to collect and model, store and retrieve, share and communicate, manipulate and analyze, present and explore geographically referenced data.

Drinking water supply is one of the most challenging issues facing the world. The latest data from the United Nations shows that water losses in water distribution networks are a serious problem, with non-revenue water, also known as unbilled water, providing water to approximately 200 million people. These issues can be addressed through digital transformation, where GIS has proven to be a valuable tool to address water management issues.

Today, decision-making is highly influenced by data. Spatial data science is a subfield of data science. It is where data science and GIS converge, with a particular focus on geospatial data and innovative computational methods. In data science, location is important in accessing, manipulating, exploring, and visualizing data through statistical computation. GIS is a fundamental information-gathering tool for successfully and effectively managing the water cycle. Thus, using and adopting a geospatial framework in combination with other tools allows for cost savings, infrastructure upgrades, quality control, and improvements in water management and water-related resources.

A spatial database can be an essential tool for successfully and effectively managing water resources. It enables the organization and management of water-related information from different sources and scales, such as sensors, satellites, studies, and models. It also allows for spatial investigation and modeling to identify designs, patterns, and relationships among water factors such as precipitation, runoff, dissipation, and invasion. The spatial database can also create and enhance water maps and charts that visualize and communicate water data and scenarios, such as water availability, demand, and allocation. It can support and aid decision-making and planning for water resources and related infrastructure, including reservoirs, dams, pipelines, and irrigation systems.

This book presents and discusses the role of computational and geospatial technologies in sustainable water resources management. Our motivation for this book is to illustrate and prove the potential of GIS and data sciences in various aspects of water resources management and sustainability. The book covers some novel developments regarding tools and technologies applied and used for water management. This book contains 14 chapters, and the key ideas of each chapter are presented successively in the three parts below:

The first part: *Bibliometry and Literature Review*
The second part: *GIS and Numerical Modeling*
The third part: *Artificial Intelligence*

Tunis, Tunisia — Noamen Rebai
Kénitra, Morocco — Aniss Moumen
Rabat, Morocco — Mohamed El Bouhaddioui

Contents

Bibliometry and Literature Review

Water Research in the Age of AI: A Bibliometric Heuristic Analysis for Trends and Opportunities

Hicham Boutracheh, Nezha Mejjad, Mohammed El Bouhadioui, and Aniss Moumen

Abstract Water is an essential resource for food, agriculture, health and industry. It is currently facing challenges posed by demographic, economic and climate change. Meanwhile, advances in artificial intelligence and machine learning (AI-ML) offer an opportunity to process large volumes of data to understand complex water-related phenomena better. This study aims to analyse the intersection of Water and AI-ML research, identifying trends shaping current research and giving insights about the future. It uses bibliometric analysis to examine Scopus data, with an incremental, heuristic approach. By the end, the analysis, based on Scival, is also crosschecked with Web of Science (WOS) data. The findings show a great dynamism in this field, with a focus on engineering approaches like "Remote Sensing," "Optical Engineering," and "Photogrammetry." Research topics emphasise detecting, measuring, modelling, and predicting hydrological, geological, and climatic phenomena. There is also a growing cross-disciplinary emphasis on energy, thermodynamics, materials, and chemistry. On the other hand, AI-ML techniques, such as "Artificial Neural Networks", "Random Forest" and "Long Short-Term Memory", are increasingly used. It also seems that the trend is towards studying parameters related to "Water Quality", "Ground Water" and "Water Levels". This particularly benefit the fields of "Agriculture" and "Water Resources Management". At the level of international collaboration, an accentuated concentration is noted with a small number of prolific countries (China, USA, India). However, industrial collaboration remains relatively weak, presenting both opportunities and challenges. Finally, this research seems to

H. Boutracheh (✉) · A. Moumen
Laboratory of Engineering Sciences, National School of Applied Sciences, Ibn Tofaïl University, Kenitra, Morocco
e-mail: hicham.boutracheh@uit.ac.ma

H. Boutracheh
IMIST, National Center for Scientific and Technical Research, Rabat, Morocco

N. Mejjad
National Centre for Nuclear Energy, Science and Technology, Kenitra, Morocco

M. El Bouhadioui
National School of Mines, Agdal, B.P. 753, Rabat, Morocco

N. Rebai et al. (eds.), *GIS, Applied Computing and Data Science for Water Management*, Lecture Notes in Geoinformation and Cartography,
https://doi.org/10.1007/978-3-031-63038-5_1

be taking a direction that is still in its infancy, due to its precocity. Nevertheless, it presents considerable potential for development, particularly in terms of international and industrial collaboration. As a perspective, the expert information comparison would certainly be beneficial to enrich this bibliometric analysis and improve the robustness of the conclusions.

Keywords Water research · WRM · Artificial intelligence · ML · Bibliometrics · Research priority · Research trends · Industrial collaboration · Patent · Scival · Incites

1 Introduction

"Water" is a central concern for humanity. It is present in the majority of human activities like agriculture, health and industry (UNESCO and World Water Assessment Programme 2023). It is also a recurring theme in religious studies, literature, poetry, music, and so forth (Brown 2015; McCarroll and Hamann 2020; Lo and Fu 2023; Oestigaard 2021; Meisch 2019). Due to accelerated changes (demographics, geopolitics, climate change, etc.), this issue is now even more critical (Kuzma et al. 2023; Boretti and Rosa 2019).

On the other hand, the revolution in information and communication technologies, culminating in the recent developments of "Artificial Intelligence" (Nasr et al. 2021; Huang and Khan 2021; Cochran et al. 2022; Walle et al. 2022), is undoubtedly a remarkable evolution (Santoro and Monin 2023; Kanzola et al. 2024). It is now possible to access and process large volumes of data to enhance understanding of phenomena and thus develop efficient solutions to complex problems (Blasch et al. 2020; Reer et al. 2023).

It is therefore useful and interesting to describe and understand how the broad field of Water has interacted with and benefited from developments in the field of "Artificial Intelligence and Machine Learning" (AI-ML). We have not found any exploratory studies in this field based on exhaustive bibliometric analysis. The studies identified focus on a particular aspect of this research (Zhao et al. 2020; Liu et al. 2023a; Ismail et al. 2023; D'amore et al. 2022; Kumari et al. 2024; Abdelmageed et al. 2022).

The objective of this study is to leverage the theoretical foundations and practical applications of bibliometrics (Richter et al. 2019; Donthu et al. 2021; Bornmann 2020) to explore research at the intersection of these two major domains, Water and AI.

2 Methodological Approach

Inspired by previous exercises (Boutracheh et al. 2023; Azirar and Boutracheh 2019; Zebakh et al. 2017), we will adopt a heuristic approach that prioritises an exploratory attitude, as suggested by the successive results of the analyses (Fig. 1).

This incremental and intuitive approach is based on the fundamental laws of bibliometrics (Hood and Wilson 2001; Wallin 2005; Gigerenzer et al. 2000; Larivière and Sugimoto 2018; Rostaing 1996), which allows for exploring large volumes of data by reducing the size of the corpora and entities to be analysed.

We propose to conduct this bibliometric study on a corpus extracted from the bibliographic database Scopus and to analyse it using the Scival analytical platform, provided by "Elsevier".

At the end of this study, an additional analysis, of some equivalent parameters, will be conducted on the Web of Science database (WOS) to verify and compare the findings on Scopus (Mongeon and Paul-Hus 2016). This will allow us to observe differently the corpus studied, notably due to disparities in coverage and scientific nomenclature between these two references.

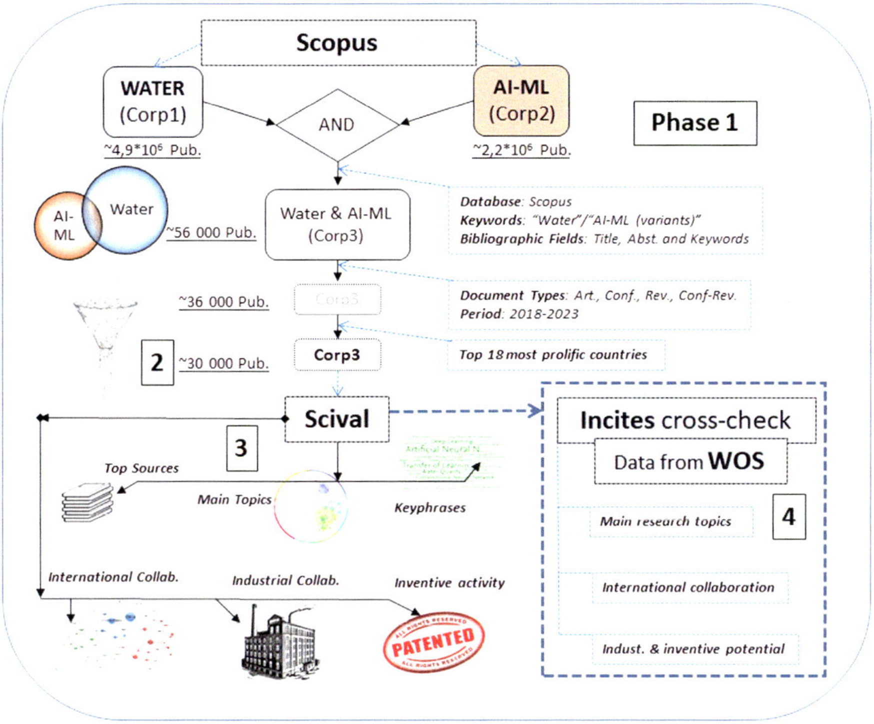

Fig. 1 Methodological approach

Figure 1 illustrates the methodological approach and presents the different stages of the study, as well as the various parameters considered in the analysis.

3 Macroscopic Overview of Scopus

3.1 Evolution Over Time

At the beginning of December 2023, the water corpus on Scopus (Corp1) stood at around 4.9 million publications, 90% of which were published after 1986. While the first publications in this corpus date back to 1788 (Fleming 1788; Hutton et al. 1784; Black 1794), research on water covers practically all major disciplinary fields. As for the AI-related corpus (Corp2), with some 2.2 million publications, it did not get off the ground until the late 1950s (Siegel and Goldstein 1959; Kersh 1958; Martens 1959; Premack 1959; Shubik 1960; Hill 1960; Minsky 1961). Over 90% of this corpus was published after 2005. It also shows a strong concentration in fields related to computer science, engineering and mathematics.

These two corpora (Water and AI) are thus at different stages of maturity and dynamism. This is reflected in the number of publications, the disciplinary profile and the shape of the evolution curve and its temporal range (Fig. 2).

The intersection of these two major fields gives rise to a corpus of around 56,000 publications (Corp3). This volume is relatively modest compared with the corpus that makes it up. It does, however, have a certain freshness, as 90% of its volume is published after 2008. So, with this promising rise in volume and a slight time lag,

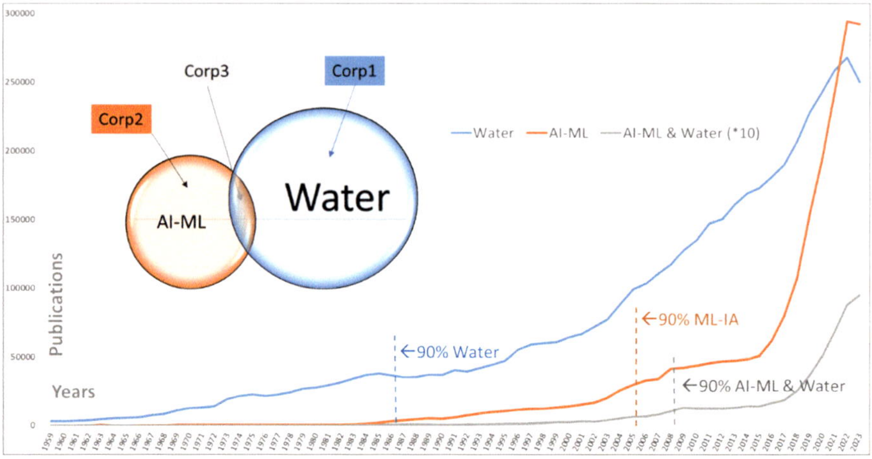

Fig. 2 Temporal evolution of publications

from 2005 to 2008, water research is embracing AI and tending to benefit from its advances.

3.2 *Broad Disciplinary Pattern*

The disciplinary proportions in the three corpora (Fig. 3) support a macroscopic analysis of the significant disciplines (~share > 5%). We can thus distinguish three main disciplinary groups:

- G1: where the disciplinary proportions (in%) of the two corpora, Corp2 and Corp3, are higher than that of Corp1 (e.g. Engineering, Computer Sces., Mathematics);
- G2: where the disciplinary proportions (in%) of the two corpora, Corp1 and Corp3, are higher than those of Corp2 (e.g. Environmental Sces, Earth and Planetary Sces, Agriculture and biological Sces., Energy, Materials, Chemistry);
- G3: made up of satellite disciplines, on the bangs of the disciplinary core of this research (e.g. "Physics and Astronomy", "Medicine").

Group G1 is thus dominated by AI's flagship disciplines. This gives an idea of the direction and nature of research at the intersection of the two fields of Water and AI (Corp3). Naturally, this could be interpreted as a dominant orientation towards integrating the possibilities of AI in the service of the focuses of Water research. This interpretation is reinforced by the importance of G2's disciplinary proportions in the intersection corpus (Corp3).

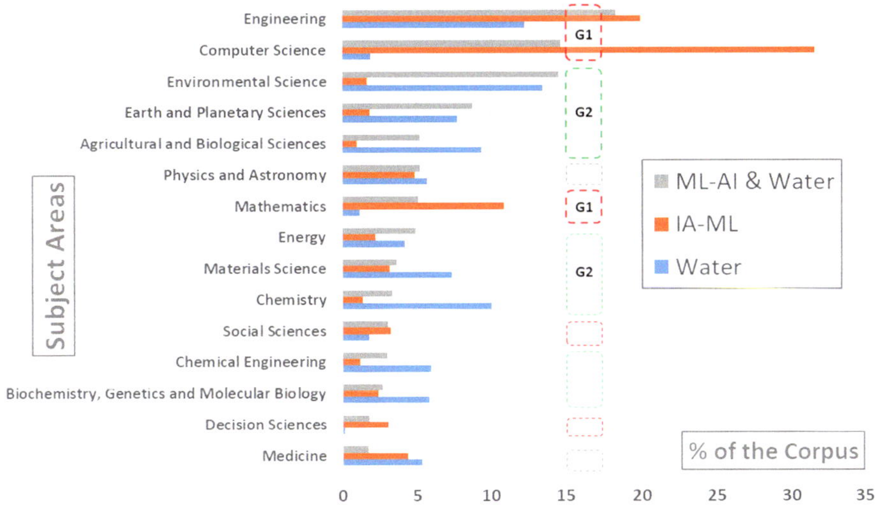

Fig. 3 General disciplinary profile

Conversely, the preponderance of primary AI disciplines in the corpus (Corp3) also shows that this intersection takes place within a logic of internal development of current AI models and practices, through the exploration of specific issues related to the Water domain.

On the other hand, the rest of the disciplinary fields, which make up group G3, show the cross-disciplinary dimension of this research and its interaction with global issues, linked in this case to Astronomy, Earth Physics, Medicine, and so on.

Finally, we have observed a certain interest in Social Sciences. This is probably due to the interaction between the disciplines of AI and the theories of "Decision" and "Behaviour". This can be explained by the use of computer science and mathematics to model and predict the phenomena under study. The latter is strongly linked to man and his ecology. This considerable aspect inspires AI applications for water issues.

3.3 Country Profiles and Collaboration Networks

For the Scopus reference, the proportion of the intersection corpus compared to that of water (Corp3/Corp1) is less than 1.2%. For the AI corpus (Corp3/Corp2), the proportion is slightly lower than 2.6%.

The shape of the Corp3 evolution curve (Fig. 2) also suggests that this is a recent and highly dynamic area of research.

Thus, following the approach proposed by Boutracheh et al. (2023), it would be highly beneficial to proceed with a selective reduction of the corpus to be examined. Thus, we will limit the analysis to publications, of the main types (Article, Conference Paper, Review and Conference Review), which were produced during the period 2018–2023. The extraction of this corpus was carried out on December 28, 2023. The result is around 36,000 publications, representing almost 65% of the initial corpus (Fig. 1). Given its dynamism, this new corpus represents a better proportion of the constituent corpora (Corp1 and Corp2), respectively 2.5 and 2.9%, compared with 1.2 and 2.6% initially.

This corpus also shows a high concentration of prolific countries. Table 1 shows the contribution of the first 18 countries, which together account for more than 82% of total publications. The top three (China, USA and India) account for over 55% of the total, with China contributing 31%.

These countries are also major partners in international collaboration in this field. The collaboration network (Fig. 4), created using the Vosviewer mapping tool (Eck and Waltman 2010), represents the 62 countries contributing more than 50 publications to this corpus. It defines three main country clusters:

- A major cluster (in blue), supported by 6 of the top 18 countries (Table 1, first column, blue cells). This group is characterized by the China-USA partnership. These are the two most prolific countries, with the highest volume of collaborations. They also mobilize the most significant partners in this corpus. Indeed,

Table 1 Prolific countries

Countries	Publications	Corpus %		
China	**11,165**	**31**	**82% of the corpus**	**55**
USA	5480	15		
India	4307	12		
Iran	2281	6		
UK	1449	4		
Sth-Korea	1338	4		
Germany	1247	3		
Canada	1245	3		
Australia	1132	3		
Saudi Arabia	1065	3		
Italy	1026	3		
Malaysia	924	3		
Japan	845	2		
Spain	843	2		
Brazil	771	2		
Turkey	763	2		
France	734	2		
Egypt	604	2		

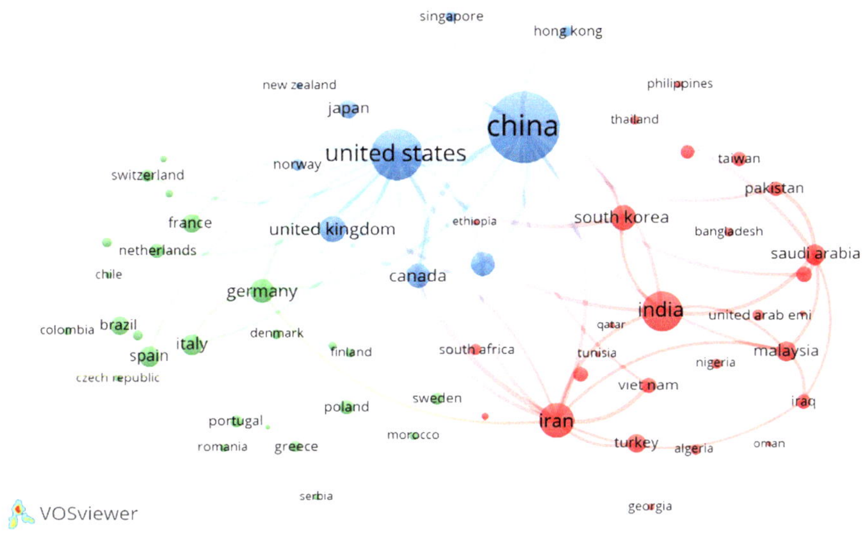

Fig. 4 Main collaborating countries (+50 publications)

beyond this cluster, China plays a major role in Asian and African collaboration. The USA, on the other hand, seems to be more dynamic at European and American levels. Canada and Australia also have strong relations with the main Asian countries. The UK, meanwhile, achieves an even balance with the rest of the world, especially Europe;
- A second group (in red), comprising 7 of the top 18 countries (Table 1, first column, red cells). India, Iran and South Korea play a central role in this cluster. In addition, India and South Korea have strong relations with China and the USA. Iran also offers greater openness to other prolific countries. Saudi Arabia and Malaysia complete the cluster's dynamic, particularly concerning Arab-Muslim and African countries;
- A third group (in green), comprising 5 of the top 18 countries (Table 1, first column, green cells). Germany and Italy are the most collaborative, establishing a structuring link between this group and the first Cluster. Germany is also developing a strong relationship with the Asian Cluster, in particular through its relationship with Iran.

4 Reducing the Corpus and Maintaining the Disciplinary Structure

Based on the theoretical foundations of bibliometrics and its potential in terms of data mining (Price 1976, 1965; de Solla Price 1978), the results of the macroscopic approach suggest that the corpus studied could be limited to the 18 most prolific countries. This reduction will improve the consistency of the data and the representativeness of the corpus, and thus enhance the efficiency of the processing and the relevance of the analyses (Donthu et al. 2021).

Together, these 18 countries account for almost 30,000 publications. This represents 77% of the corpus studied, which concentrates on the core of the distribution and general dynamics of this field. We will therefore carry out an additional verification at the level of the disciplines covered.

From a disciplinary point of view, the table (Table 2) represents the main disciplines of the reduced corpus (+1100 Pubs/+4% of the corpus). The "Engineering" and "Environmental Science" fields structure the corpus. Together, they account for almost 60% of the total, with a sharp increase over the last two years. The "Computer Sciences" and "Earth and Planetary Sciences" fields are next most important, in terms of both volume and growth. Other less prolific fields, "Physics and Astronomy", "Agriculture and Biological Sciences", "Energy" and "Mathematics", showed remarkable growth during the period. The rest of the disciplines, particularly those related to chemistry, are relatively modest but significant, particularly because of the multidisciplinarity (Calver et al. 2018; Gagné et al. 2018) that can exist between the different disciplinary fields.

To sum up, the first 4 disciplinary fields in the table make up almost 82% of the overall corpus. The top 10 make up nearly 95%. This confirms the structuring

Table 2 Stability of the disciplinary structure

Subject areas/Years	2018	2019	2020	2021	2022	2023	Corpus %
Engineering	688	979	1349	1732	2531	2868	**35**
Environmental science	608	797	1150	1653	2122	2788	**31**
Computer science	609	930	1121	1422	1861	1953	**27**
Earth and planetary sciences	349	539	808	1123	1416	1544	**20**
Physics and astronomy	220	337	531	661	847	894	**12**
Agricultural and biological sciences	253	353	467	649	781	981	**12**
Energy	207	273	437	516	785	839	**10**
Mathematics	188	343	379	467	664	735	**9**
Materials science	153	222	349	402	592	723	**8**
Chemistry	136	203	305	389	561	659	**8**
Social Sciences	129	197	280	349	489	605	**7**
Bioch-genetics and mol-biology	126	160	261	334	470	558	**6**
Chemical engineering	123	162	225	328	461	571	**6**
Medicine	68	132	140	235	337	353	**4**
Decision sciences	67	126	151	231	308	286	**4**

importance of the top of the list and their interaction with the rest of the disciplines in this corpus. This further reinforces and clarifies the first analysis (Fig. 3) and also gives an idea of disciplinary concentration and interdependence. This reveals, in passing, the stability of the disciplinary structure and thus supports the relevance of the incremental bibliometric approach. We will therefore retain the reduced corpus (Corp3) for further analysis.

5 Deepening with Incremental Bibliometric Approach

We transfer the reduced Corp3 data to the Scival analytical platform (Dresbeck 2015; Cucari et al. 2023) for further analysis (Fig. 1). This corpus, made up of 29,260 publications and involving nearly 104,000 researchers, receives an average of nearly 11.7 citations per publication and obtains an average "Field-Weighted Citation Impact" (FWCI[1]) value of 1.58 (Purkayastha et al. 2019).

Thus, FWCI equal 1.58 means that, on average, the publications in this group have a citation impact 58% higher than the international average. This confirms the relevance of this research theme.

[1] FWCI is a standardized indicator of the average impact of a set of publications. It takes into account the context of citations and the type, year and field of each publication in the corpus. This makes it possible to calculate a normalized value for a publication, or a corpus, in proportion to the entire database.

Indeed, almost 54% of the publications are in the first quartile journals (Q1), as defined by the "Source Normalized Impact per Paper" (SNIP[2]) (Waltman et al. 2013; Brown and Gutman 2019). Nearly 79% of the corpus is also in the first two quartiles (Q1 and Q2) and less than 9% in the last quartile (Q4).

Moreover, 18.5% of publications belong to the top 10% worldwide in terms of FWCI. At the same time, 24.9% of the corpus is published in the world's top 10% of journals in terms of SNIP.

These relatively high figures confirm the relevance of this theme. In addition, the difference between these two proportions (18.5 and 24.9%), which can be considered relatively small, could mean that the orientation of "high-performance journals" to cover this theme is well "rewarded" by the interest of the scientific community that reads and cites this research.

5.1 Insights from the Main Publication Sources

Among the top 100 most prolific sources (journals, proceedings), Table 3 focuses on the first ones on the list, then those with the best SNIP and finally those in which the corpus generates the best FWCI. Analysis of the scopes of these journals suggests an exploratory approach to the main research orientations of the field (Meller et al. 2023; Duguet et al. 2024):

- **A clear orientation towards the use of knowledge and application engineering approaches**, from the fields of "Remote Sensing", "Optical Engineering", "Sensors", "Chemicals", "Instrumentation", "Photogrammetry", "Materials", "Liquids" and so on.
- **And to exploit the data,** resolutely accessible and voluminous, resulting from these observation and characterization techniques, **by drawing on the new opportunities offered b**y "Computers and Electronics", "Artificial Intelligence", "Neural Computing", "Informatics", "Networks and Systems", "Geoinformation", "Communication", etc.
- **To observe and understand phenomena related to** "Water Sciences", in the sense of "Environment", "Hydrology", "Oceans", "Geoscience", "Earth", "Chemosphere", etc.
- **For "Measurement",** "Assessment", "Management", "Monitoring" **of issues linked to** "Pollution", "Sustainability", "Cleaner Production", "Agriculture", "Agronomy", "Bioresource", "Energy", "Fuel", "Petroleum", "Construction and Building Materials", etc.

This finding, based on a heuristic approach to analysing the areas of interest of the main sources, revealed a direction that is consistent with the previous results. That of

[2] The SNIP metric is a measure of the citation impact for Scopus journals. It takes into account differences in citation practices between fields. This makes it possible to compare journals from different fields.

Table 3 Main sources by "share of publications", SNIP and FWCI

Top sources (Pub. > 124)	Pub.	Top sources (SNIP > 1.7)	Pub.	SNIP	Top corpus (FWCI > 2.1)	Pub.	FWCI
"Remote Sensing"	867	"IEEE Transactions on Geoscience and Remote Sensing"	165	3.61	"Remote Sensing of Environment"	140	4.31
"Water (Switzerland)"	722	"Remote Sensing of Environment"	140	3.53	"Nature Communications"	71	3.57
"Journal of Hydrology"	469	"ISPRS Journal of Photogrammetry and Remote Sensing"	50	3.28	"Journal of Hydrology"	469	3.51
"Science of the Total Environment"	414	"Nature Communications"	71	3.27	"Construction and Building Materials"	92	3.3
"IEEE Access"	280	"Applied Energy"	49	2.76	"Journal of Molecular Liquids"	58	3.18
"Proceedings of SPIE—The International Society for Optical Engineering"	270	"Computers and Electronics in Agriculture"	186	2.47	"Computers and Electronics in Agriculture"	186	3.11
"Sensors"	263	"Water Research"	158	2.42	"Measurement: Journal of the International Measurement Confederation"	45	3.08
"Environmental Science and Pollution Research"	250	"Journal of Cleaner Production"	158	2.38	"Chemical Engineering Journal"	46	3.05
"Sustainability"	242	"IEEE Geoscience and Remote Sensing Letters"	74	2.36	"Agricultural Water Management"	128	3.02
"Journal of Physics: Conference Series"	235	"Engineering Applications of Artificial Intelligence"	46	2.27	"Hydrology and Earth System Sciences"	64	3.02
"Water Resources Management"	226	"Construction and Building Materials"	92	2.21	"Environmental Research"	73	3.01
"Scientific Reports"	205	"Chemical Engineering Journal"	46	2.18	"Applied Energy"	49	2.86

(continued)

Table 3 (continued)

Top sources (Pub. > 124)	Pub.	Top sources (SNIP > 1.7)	Pub.	SNIP	Top corpus (FWCI > 2.1)	Pub.	FWCI
"International Geoscience and Remote Sensing Symposium (IGARSS)"	196	"Journal of Hazardous Materials"	62	2.14	"Process Safety and Environmental Protection"	62	2.86
"Journal of Environmental Management"	190	"Energy"	68	2.13	"Applied Water Science"	68	2.78
"Computers and Electronics in Agriculture"	186	"Ocean Engineering"	157	2.11	"Chemosphere"	134	2.71
"Applied Sciences (Switzerland)"	185	"Environmental Science and Technology"	101	2.05	"Journal of Hazardous Materials"	62	2.67
"Lecture Notes in Computer Science"	177	"Science of the Total Environment"	414	2.03	"Oceans Conference Record (IEEE)"	47	2.67
"IEEE Transactions on Geoscience and Remote Sensing"	165	"Agricultural Water Management"	128	2.02	"ISPRS Journal of Photogrammetry and Remote Sensing"	50	2.62
"Journal of Cleaner Production"	158	"Bioresource Technology"	47	2.02	"Fuel"	45	2.61
"Water Research"	158	"Applied Water Science"	68	1.93	"Ecological Informatics"	65	2.61
"Ocean Engineering"	157	"Fuel"	45	1.9	"Ecological Indicators"	83	2.55
"Lecture Notes in Electrical Engineering"	153	"Applied Thermal Engineering"	50	1.88	"Materials"	64	2.45
"Water Science and Technology: Water Supply"	142	"Process Safety and Environmental Protection"	62	1.85	"Journal of Cleaner Production"	158	2.4
"Remote Sensing of Environment"	140	"Journal of Environmental Management"	190	1.85	"Energy"	68	2.34
"ACM International Conference Proceeding Series"	140	"International Journal of Applied Earth Observation and Geoinformation"	70	1.83	"Water Resources Research"	118	2.34

(continued)

Table 3 (continued)

Top sources (Pub. > 124)	Pub.	Top sources (SNIP > 1.7)	Pub.	SNIP	Top corpus (FWCI > 2.1)	Pub.	FWCI
"Chemosphere"	134	"Neural Computing and Applications"	122	1.82	"Agronomy"	56	2.25
"Lecture Notes in Networks and Systems"	133	"Measurement: Journal of the International Measurement Confederation"	45	1.78	"Journal of Chemical Physics"	72	2.18
"Agricultural Water Management"	128	"IEEE Journal of Selected Topics in Applied Earth Observations and Remote Sensing"	125	1.76	"Bioresource Technology"	47	2.16
"IOP Conference Series: Earth and Environmental Science"	127	"Journal of Petroleum Science and Engineering"	83	1.74	"International Journal of Applied Earth Observation and Geoinformation"	70	2.16
"Environmental Monitoring and Assessment"	127	"Journal of Hydrology"	469	1.73	"Water Research"	158	2.14
"IEEE Journal of Selected Topics in Applied Earth Observations and Remote Sensing"	125	"IEEE Transactions on Instrumentation and Measurement"	46	1.72	"Engineering Applications of Artificial Intelligence"	46	2.13

adopting AI, and its advances, in the development of observation and analysis techniques for water-related phenomena and applications. Some conventional purposes, linked for example to sustainability and agriculture, are accompanied by other fields of interest such as energy, oil and construction materials. This has enabled us to develop a kind of analysis that we will attempt to elucidate in the following steps.

This analysis considered a subset of publication sources, which we took to be representative of researchers' practices in this field. This has enabled us to develop a certain reading that needs to be verified by exploring complementary parameters.

5.2 *Key Topics in Water-AI Research*

Scival provides an analysis of research themes or "Topics". A Topic is a group of publications focusing on a single research question. There are ~ 96,000 Topics, grouped into around 1500 clusters (Elsevier 2023).

Topics are also dynamic and disjoint, formed using a method based on the density of the citation network. They are characterized by a (Prominence) index, which represents the current "momentum" of each research item. Then, each Topic is ranked on a scale of 100, to obtain its "Prominence percentile", which reflects its relative interest in the topic space (Mokhnacheva and Tsvetkova 2021). This makes it possible to support more bibliographic analyses to describe major research orientations and opportunities (Wen et al. 2020).

In this case, the publications in our corpus are present in 4904 of the total 96,000 topics. Our analysis will focus on topics in—the top 1% (Prominence Percentile)—where corpus participation is prolific—with positive evolution and high impact (FWCI). This gives a list of almost 43 topics, out of 4904, presented in Table 4:

- Topics in which corpus participation exceeds 3%. This group of eight topics is characterised by a total number of publications exceeding 95 each. The Topic "Prediction; Flood Forecasting; Water Tables", with its 2,448 publications, its 40.95% participation and its 38.9% growth (2018–2022), is undoubtedly an outstanding topic in the corpus. The themes, like "Radar Altimetry; Flood Hazards; Remote Sensing", "Crops; Leaf Area Index; Hyperspectral Data" and "Soil Color; Near-Infrared Spectroscopy; Hyperspectral", are also showing remarkable participation and growth;
- Topics in which participation is between 2 and 3%. This group of eight topics is characterised by a volume of publications exceeding 63 each. The "Landslides; Debris Flow; Susceptibility" topic is the most prolific, with a growth rate of 106.3% between 2018 and 2022. The topics "Underwater; Haze; Atmospheric Scattering", "River Basin; Nonpoint Source Pollution; Stream Flow" and "Air Conditioning; Building Controls; Thermal Comfort" have remarkable growth rates of 530.4, 240.6 and 320% respectively.
- Topics in which participation is between 0.9% and 2%. This group of 11 topics is characterised by the prolific case of "Object Detection; Deep Learning; IOU", with 1274 publications and a growth rate of almost 80%. Nevertheless, the contribution to this topic does not exceed 0.96%. On the other hand, the topics "Deep Learning; Handwriting Recognition; Long Short-Term Memory", "Climate Change; Glacial Lake; Ice Cover", "Brines; Carbon Sequestration; Carbon Dioxide" and "Wastewater Treatment; Anaerobic Digesters; Fouling", with contributions exceeding 1.17%, exhibit high evolution rates of 348.4, 579.9, 611.5 and 546.2% respectively.
- Topics with less than 0.9% participation. This group of 16 topics compensates for low contributions with high evolution rates. The first three most prolific topics, "Plastics; Marine Debris; Litter", "Biosorbents; Congo Red; Adsorption" and "Fog Computing; Network Protocol; Internet of Things", participate with 116, 87 and 67 respectively, in terms of publications, and 378.6, 81 and 315.9% as evolution rates.

Table 4 Main topics of the water-AI research

Topic's name	Pubs.	Topic share (%)	Growth 2018–22 (%)	FWCI
"Prediction; Flood Forecasting; Water Tables"	2448	40.95	38.9	2.06
"Object Detection; Deep Learning; IOU"	1274	0.96	79.9	1.58
"Radar Altimetry; Flood Hazards; Remote Sensing"	365	7.99	141.9	1.52
"Heat Transfer; Heat Transfer Enhancement; Automobile Radiators"	289	3.2	71	2.81
"Landsat; Land Cover; Remote Sensing"	254	3.53	65.2	2.21
"Crops; Leaf Area Index; Hyperspectral Data"	226	4.26	180.3	1.99
"Landslides; Debris Flow; Susceptibility"	210	2.78	106.3	2.75
"Underwater; Haze; Atmospheric Scattering"	174	2.8	530.4	1.95
"Droughts; Evapotranspiration; China"	170	3.24	141.2	2.04
"TRMM; China; Rain Gages"	140	4.93	110.3	1.68
"Plastics; Marine Debris; Litter"	116	0.57	378.6	1.85
"Sleuths; Cellular Automaton Model; Urbanization"	106	2.45	72	1.87
"Soil Color; Near-Infrared Spectroscopy; Hyperspectral"	95	3.23	141.6	1.27
"Groundwater Samples; Water Quality; Piper"	92	1.79	68.8	3.33
"Deep Learning; Handwriting Recognition; Long Short-Term Memory"	90	1.37	348.4	2.01
"River Basin; Nonpoint Source Pollution; Stream Flow"	89	2.46	240.6	1.24
"Biosorbents; Congo Red; Adsorption"	87	0.57	81	2.94
"Climate Change; CMIP; Regional Climate"	86	2.06	127.3	1.19

(continued)

Table 4 (continued)

Topic's name	Pubs.	Topic share (%)	Growth 2018–22 (%)	FWCI
"Spatial Autocorrelation; Soil Organic Carbon; Carbon Sinks"	68	2.75	101.6	2.35
"Water Footprint; Water-Energy Nexus; Nexus"	68	1.28	1.1	1.98
"Fog Computing; Network Protocol; Internet of Things"	67	0.45	315.9	2.57
"Solar Still; Humidifiers; Desalination"	64	2.19	204.9	3.82
"Air Conditioning; Building Controls; Thermal Comfort"	63	2.07	320	1.45
"Attack; Supervisory Control; Intrusion Detection"	61	1.61	33	2.6
"Concrete Aggregates; Constr-Demolition Waste; Mechan. Properties"	60	1.07	79.2	3.57
"Crops; CERES (Experiment); Climate Change Impact"	56	0.79	403.6	1.76
"PPCP; Micropollutant; Drugs"	45	0.5	290	1.29
"Networked Control Systems; Denial-Of-Service Attack; Attack"	43	1.24	286.9	1.16
"Geopolymers; Coal Ash; Alkali"	42	0.34	518.1	2.88
"Climate Change; Glacial Lake; Ice Cover"	40	1.17	579.9	0.95
"Smart Cities; Big Data; Internet of Things"	38	0.41	502.8	1.85
"Mechanical Properties; Self Compacting Concrete; Concrete Slabs"	36	0.61	468.7	2.8
"Microcystin; Algal Bloom; Cyanobacteria"	35	1.32	232	1.36
"Demand Response; Dem. Side Management; Energy Conservation"	34	0.25	530.2	2.03
"Hot Temperature; Phase Change Materials; Photovoltaic System"	34	0.78	258.7	2.6
"Insar; Monitoring; Subsidence"	33	0.77	489.2	1.6

(continued)

Table 4 (continued)

Topic's name	Pubs.	Topic share (%)	Growth 2018–22 (%)	FWCI
"Brines; Carbon Sequestration; Carbon Dioxide"	32	1.2	611.5	1.86
"Wastewater Treatment; Anaerobic Digesters; Fouling"	30	1.49	546.2	1.62
"Useful Life; Health Care Management; Rolling Bearing"	26	0.46	304.9	1.19
"A-intelligence; Travel Time; Advanced Traveler Information Systems"	24	0.39	228.3	1.62
"Electrochemical Oxidation; Diffusion Electrodes; Degradation"	21	0.46	212.3	1.52
"Intrusion Detection Syst.; Network Security; Denial-Of-Service Attack"	21	0.21	1.1	3.19
"Cement; Rice Hulls; Mortar"	20	0.82	248.6	2.22

5.3 *Main Trends Arising from the Topics*

In other words, the analysis of the most representative topics corroborates the analysis of the main scientific journals, with additional elements of precision.

First, a high concentration of publications is observed for the two topics "Prediction; Flood Forecasting; Water Tables" (Chen et al. 2023; Uddin et al. 2023; Rawat et al. 2023) and "Object Detection; Deep Learning; IOU" (Guru Prasad et al. 2023; Ting et al. 2021; Bawankule et al. 2023). The third topic, "Radar Altimetry; Flood Hazards; Remote Sensing", has grown by almost 142%, and deals with the detection and measurement of hydric, geological and climatic phenomena, for modelling, prediction and anticipation purposes, to better manage and monitor these phenomena and their ecological impacts (Tariq and Qin 1985; Abdalla et al. 2021; Chen et al. 2018).

Second, a smaller group of topics, ranging from 67 to 226 publications, with high growth rates, from 140 to 530%, represent some of these areas of interest. Thus, there are topics reflecting issues linked to climate, agriculture, and pollution, like "Crops; Leaf Area Index; Hyperspectral Data" (Peng and Karimi Sadaghiani 2023), "Droughts; Evapotranspiration; China" (Jiang et al. 2023), "Plastics; Marine Debris; Litter" (Min et al. 2020), "River Basin; Nonpoint Source Pollution; Stream Flow" (Wang et al. 2022). Others are related to data acquisition, treatment and interpretation, such as "Soil Color; Near-Infrared Spectroscopy; Hyperspectral" (Xiao 2023), "Deep

Learning; Handwriting Recognition; Long Short-Term Memory" (Wei et al. 2021), "Fog Computing; Network Protocol; Internet of Things" (Kasiselvanathan et al. 2023).

Finally, other less prolific but very dynamic topics, such as "Air Conditioning; Building Controls; Thermal Comfort" (Fang 2022), "Crops; CERES; Climate Change Impact" (Batool et al. 2022), "Geopolymers; Coal Ash; Alkali" (Shahmansouri et al. 2021), "Climate Change; Glacial Lake; Ice Cover" (Bolibar et al. 2022), "Demand Response; Demand Side Management; Energy Conservation" (Xu et al. 2021), "Brines; Carbon Sequestration; Carbon Dioxide" (Mohammadian et al. 2023), "Wastewater Treatment; Anaerobic Digesters; Fouling" (Wijaya and Oh 2023), all reveal cross-cutting aspects relating to energy, thermodynamics, materials and chemistry.

Indeed, as in the case of "Heat Transfer; Heat Transfer Enhancement; Automobile Radiators" (Kanti et al. 2023; Giwa et al. 2021; Vicki Wanatasanappan et al. 2023), the dynamism of this group of topics, translated by remarkable evolution rates, shows that the research (water and AI) presents an applicative interest for these major disciplinary fields and that there is great potential for exchanges and mutual transfers at this crossroads.

Figure 5 summarises the previous analyses using a wheel view (Scival). It presents a mapping of the main topics (top 1%) involved in the corpus studied. It shows the proximity of the most significant topics and their concentration in the major disciplinary fields. This synoptic analysis is also consistent with the exploratory scan of the most relevant scientific journals.

In fact, from the "Earth/Environment" cluster, represented by T.1472, to the "Computer/Mathematics" cluster, represented by T.4338, the corpus studied addresses interdependent issues, ranging from engineering, agriculture and biochemistry to decision sciences, with intense activity in energy, chemistry and materials.

5.4 *Terminology Analysis: "Keyphrases"*

To deepen the previous analyses, we are exploring the most relevant keywords used in this research. As suggested by Boutracheh et al. (2024), keyword analysis is a complementary dimension that allows us to identify specific concepts addressed in the publications.

To identify relevant keywords, Scival uses its "Fingerprint" engine (Elsevier 2024) and applies several text mining and NPL algorithms to the bibliographic unstructured information (Title, Abstracts, keywords, etc.). Each publication is assigned standardised keyphrases, with a weight that reflects its relative importance within the corpus studied. Thus, at the level of a given disciplinary field, topic or corpus, the keywords (keyphrases) are given a value, between 0 and 1, which reflects its general interest (relevance) in this corpus. The value of 1 is the maximum.

For the corpus studied, we thus obtain a word cloud (Fig. 6) which represents the top 50 relevant keyphrases, the size reflecting the relevance (0–1). In brief, this group

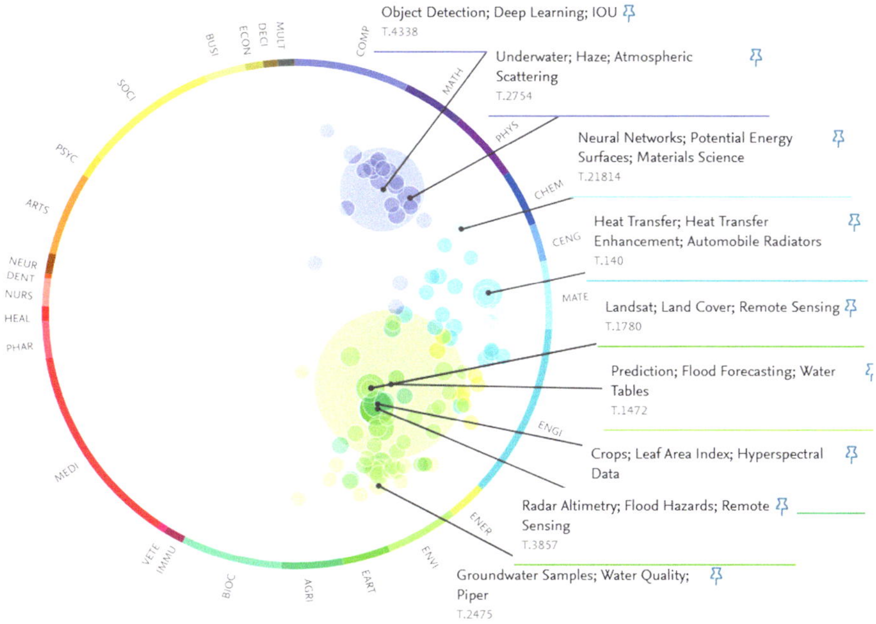

Fig. 5 Main topics in the disciplinary wheel

of keyphrases is present in almost 87% of the publications in the total corpus. This provides a good representation of the concepts covered. In addition, all these terms show a positive trend (green colour) over the period 2028–2022. Almost two-thirds of the terms show a growth rate of over 200%, from a minimum of 38.9% for "Extreme Learning Machine" to a maximum of 2215% for "Long Short-term Memory". This finding is totally in line with the dynamism and general trend of the corpus studied.

Furthermore, the concepts of AI-ML, in particular all that revolves around the "Neural Network" and "Machine Learning", and their algorithms, seem to dominate the cloud of keyphrases, both in terms of occurrence (n/50) and in terms of relevance (size). These terms are also the most common in the corpus. Specifically, "Artificial Neural Network" (Raissi et al. 2019) and "Transfer of Learning" (Ghosh et al. 2023) are the most relevant, with scores of 1 and 0.78. They are also the most frequent, with 8052 and 9904 publications concerned. In terms of dynamism, "Long Short-term Memory" (Li et al. 2019a), present in 1198 publications and with a score of 0.33, has a spectacular growth rate of 2215.8%. While "Random Forest", present in 2743 publications and scoring 0.35, has a growth rate of 560.2%.

On the other hand, of the 50 keyphrases in the cloud, 21 are directly linked to "Water". The three most relevant (0.64, 0.31 and 0.3), "Water Quality" (Ghosh et al. 2023), "Groundwater" (Ruidas et al. 2023) and "Water Levels" (Zhang et al. 2023), are involved in 2687, 1486 and 1066 publications respectively. They also show a growth rate of 248.1, 309.9 and 246.9%.

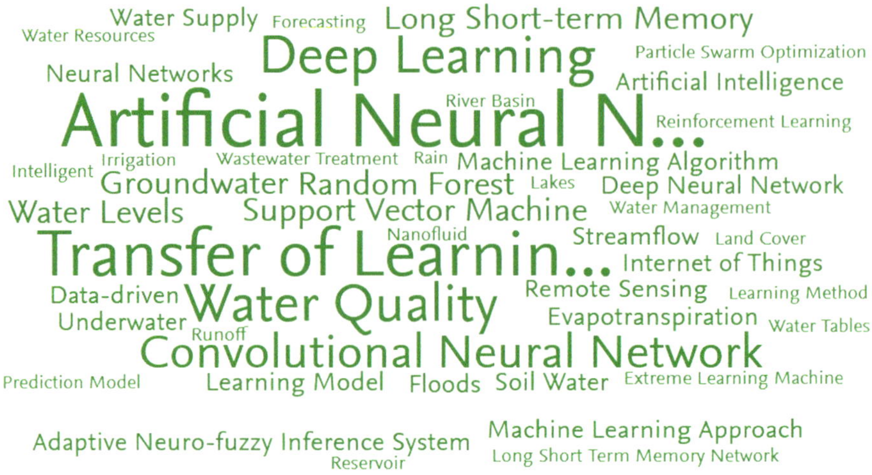

Fig. 6 Top 50 relevant keyphrases of the general corpus

In terms of concepts, the issue of "water resource management" is highly present in this group. It is closely linked to the keyphrases of "Wastewater", "Water/Resources, Quality, Levels, Supply, Management" (Li et al. 2020a; Navya and Sudha 2523) and "Rain, Floods, Lakes, River Bassin, Runoff".

Another related aspect concerns the observation of phenomena linked to water and land, such as "Land Cover", "Ground, Soil, Under/Water", "Water Tables" and "Streamflow" (Adnan et al. 2023). This observation issue is being addressed, particularly using "Remote Sensing" techniques, a key phrase present in 2420 publications, with an overall increase of 372.3%.

Observation data relating to these priority issues are processed by AI-ML methodologies. Due to their high development rates, these techniques appear to be in a state of adaptation to water-related issues.

The ultimate purpose of these developments in data processing is "forecasting", a keyphrase present in 4380 publications, with an evolving rate of 219.4%. This makes it possible to draw up projections to predict and better anticipate the phenomena studied (Ghosh et al. 2023; Elsheikh et al. 2023). This also benefits the main areas concerned, such as agriculture (Gaagai et al. 2023; Nunno and Granata 2023) and water-related techniques and materials (Nazar et al. 2023), or even niche areas such as "Nanofluid" (Vicki Wanatasanappan et al. 2023), a key phrase present in 336 publications, with a 172% increase.

5.5 Two Cases of More In-Depth Terminology Analysis

In the previous analysis of topics, we considered the 43 most relevant topics, with a limitation to the top 1% in terms of prominence. We then explored the 50 most relevant keyphrases in the corpus, with no restriction on topics. We are now taking a closer look at two representative cases, particularly in terms of size and the proportion to the topic. Without being exhaustive, this will illustrate our approach and will enrich the first results in the process. We illustrate the following two cases:

1. Set1: "MERIS; Water Quality; Suspended Matter" (T.3333). This is the third most prolific topic in the corpus. It is present in 431 publications, with a 20.1% contribution to this topic and an increase of 132.6%, during the 2018–22 period;
2. Set2: "Evapotranspiration; Crop Coefficient; Artificial Neural Network" (T.4843), present in 333 publications, with a share of 22.97% and an evolution of 122.7%.

(1) **Set1: "MERIS; Water Quality; Suspended Matter"**

This Set uses the term "MERIS", which represents a spectroscopic imaging technique. The cloud of Keyphrases that represents it (Fig. 7) provides a more detailed terminology than that of the entire corpus. This set is focused on the use of spatial imaging techniques (Landsat, "Remote Sensing") for the main purpose of measuring, assessing and monitoring "Water Quality", particularly in "Lakes", "Estuaries", "Coastal Water", etc.

The "Quality Parameters" are approached using new data analysis techniques derived from AI. This approach appears to be linked to parameters that are showing clear signs of progress, such as "Chlorophyll", "Algal Bloom", "Cyanobacteria",

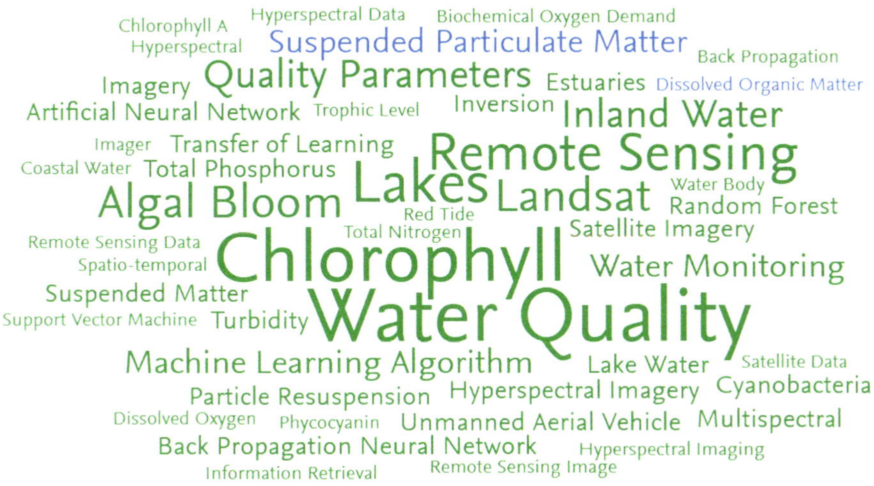

Fig. 7 Top 50 relevant keyphrases of the Set1 topic

"Total Phosphorus", "Total Nitrogen", "Turbidity", "Dissolved Oxygen", "Red Tide", and so on (Pahlevan 2020; Pyo 2019; Davis et al. 2023; Liu et al. 2022). On the other hand, with the blue colour in the word cloud, there has been a downturn in interest in "Suspended Particle Matter" and "Dissolved Organic Matter". It is very likely that the new imaging techniques, combined with advances in AI-ML algorithms, will give meaning to this observation and enable us to explore techniques and parameters for measuring "water quality" that were not previously economically available.

(2) **Set2: "Evapotranspiration; Crop Coef.; Artificial Neural Network"**

The word cloud for this Set (Fig. 8) is centred on "Evapotranspiration", a Keyphrase with the maximal relevance of 1. It is present in 65% of the publications on this topic and will grow by 350% between 2018 and 2022. "Evapotranspiration" is the natural process by which water is lost through evaporation, particularly in crops. This phenomenon is critical in agriculture, particularly in Arid/Semiarid Regions. So, for better "Water resources management" in these regions, it is necessary to collect and process a huge amount of data, particularly "Climate and Meteorological Data" (Mostafa et al. 2023), to "Estimate" parameters such as "Crop Coefficient" (Ikram et al. 2023) and "Soil Water", to be able to model this phenomenon and better manage "Water resources" (Katimbo 2023).

This approach benefits from AI-ML algorithms such as the "Artificial Neural Network" and the "Long Short Term Memory Network". On the other hand, other approaches (shown in blue) seem to be losing interest, such as the "Genetic Algorithm" and the "Back Propagation Neural Network".

In sum, it appears that this incremental approach has the merit of highlighting certain aspects that were not clear from the first approaches. It would therefore

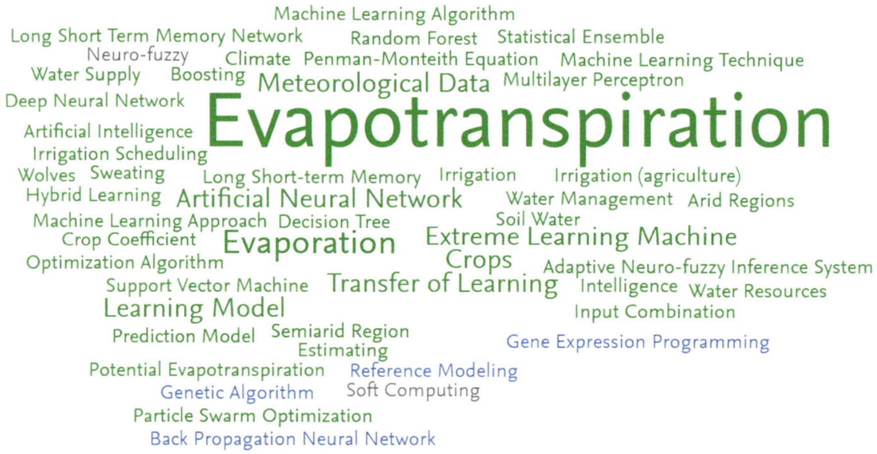

Fig. 8 Top 50 relevant keyphrases of the Set2 topic

be interesting to multiply these exploratory cases to enrich our knowledge on this subject.

6 Different Aspects of Collaboration

The total corpus studied achieved an FWCI of 1.58. However, the international collaboration sub-corpus, which represents 29.8% of the total, does much better, with an FWCI of 2.26 and almost 90% of its publications are in Q1 and Q2 journals (SNIP). More than 28% of these collaborations are in the top 10% of publications (FWCI) and almost 46% are published in the top 10% of journals (SNIP). On the other hand, industrial collaboration, at only 3.5% of the overall corpus, has an FWCI of 1.59, which is almost the same as the general performance, but lower than what is expected for industrial research (Perkmann and Walsh 2009).

In addition to this observation, the recent dynamism of the corpus studied suggests that this theme, which tends to be internationalised, is still at an early stage of scientific development and is not yet sufficiently mature from an industrial point of view. It is also a very broad field, linked to universal issues such as climate, earth, ocean, agriculture, remote sensing and so on. This encourages collaboration and international projects, mainly involving the world of academics.

Exploring this international collaboration would provide greater precision in the structuring orientations of this theme. It might also help to explain the lack of interest shown by industry in this area.

6.1 The Major Disciplines of International Collaboration

All the publications in Scopus are indexed, at least, in one major discipline "Subject Area", from a list of 26. The collaboration corpus studied shows the disciplinary distribution detailed in Fig. 9. Of the 26 major subject areas, 5 contributed more than 1140 publications. Except for "Agricultural and Biological Sciences", which contributes 13.1% to the corpus, the other four contribute more than 20%. Nevertheless, it is essential to explore the sub-disciplinary fields (Subcategories) to deepen this analysis.

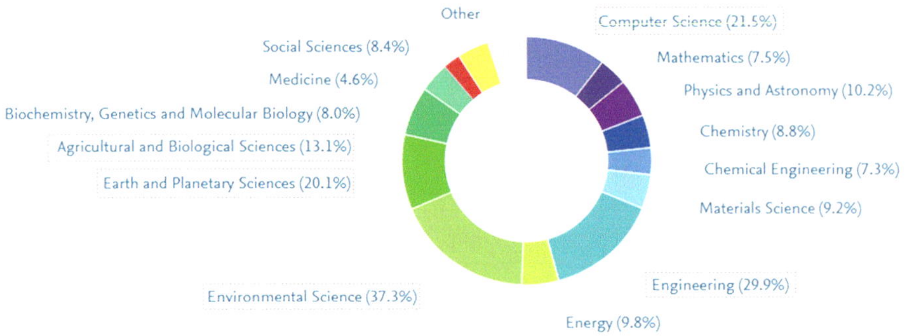

Fig. 9 Major subject areas of the international collaboration

6.2 International Collaboration: General Trend

Table 5 lists the 19 Subcategories that contribute most (+4.5%) to the overall corpus of collaboration. This group is dominated by the Subcategories belonging to the Subject Area "Environmental Science", with 7 Subcategories, including 4 in the Top 5 and the 2 most prolific on the list: "Water Science and Technology" and "Environmental Engineering", with a contribution of 16.2 and 9.5% respectively to the corpus of international collaboration (Ruidas et al. 2023; Liu et al. 2023b; Khosravi et al. 2023).

Table 5 Main prolific subcategories of the International Collaboration

Subject area	Subcategory	% of corpus	FWCI
Environmental science	**"Water Science and Technology"**	**16.2**	**2.4**
Environmental science	**"Environmental Engineering"**	**9.5**	**2.24**
Computer science	**"Computer Science Applications"**	**8.6**	**2.01**
Environmental science	**"Pollution"**	**8.5**	**2.16**
Environmental science	**"Environmental Chemistry"**	**8.5**	**2.17**
Engineering	**"Electrical and Electronic Engineering"**	**8.1**	**1.94**
Social sciences	**"Geography, Planning and Development"**	**7.1**	**2.3**
Earth and planetary sciences	**"General Earth and Planetary Sciences"**	**7.0**	**2.01**
Energy	**"Renewable Energy, Sustainability and the Environment"**	**6.5**	**2.2**
Engineering	**"Civil and Structural Engineering"**	**6.3**	**2.42**
Environmental science	**"Waste Management and Disposal"**	**6.0**	**2.16**
Materials science	**"General Materials Science"**	**5.8**	**2.45**

(continued)

Table 5 (continued)

Subject area	Subcategory	% of corpus	FWCI
Environmental science	**"Management, Monitoring, Policy and Law"**	**5.4**	**2.13**
Energy	**"Energy Engineering and Power Technology"**	**5.3**	**2**
Bioch., gene., and mol. biology	**"Biochemistry"**	**5.3**	**2.38**
Environmental science	**"General Environmental Science"**	**4.9**	**2.12**
Computer science	**"Artificial Intelligence"**	**4.8**	**1.93**
Agricultural and biolog. sc.	**"Aquatic Science"**	**4.5**	**2.7**
Computer science	**"Computer Networks and Communications"**	**4.5**	**1.66**

Next came "Computer Science", with 3 contributions. Then come "Engineering" and "Energy", with 2 entries each. "Social Sciences" follows, with its subcategory "Geography, Planning and Development". Finally, the Subject Areas with just one participation each are "Earth and Planetary Sciences", "Materials Science", "Biochemistry, Genetics and Molecular Biology" as well as "Agricultural and Biological Sciences", with its Subcategory "Aquatic Science" achieving the highest FWCI on the list (2.7).

To sum up, there is one structuring trend in international collaboration that can be seen in this table, which is also very much in line with the general trend. This involves the application of advances in computer sciences, AI and communication technologies to developing water sciences and technologies. Issues relating to the observation of the environment, the earth and the aquatic world are addressed to understand and manage phenomena related to water and climate ("pollution", "waste*", "energy"). This is undertaken with a focus on ecology and sustainability ("Waste Management and Disposal", "Renewable Energy", "Sustainability and the Environment"). Naturally, this brings to mind techniques for observing ecosystems ("Geography", "Earth and Planetary Sciences") and measuring elements ("Environmental Chemistry", "Biochemistry", "Electrical and Electronic Engineering", "Environmental Engineering"). Finally, all this is related to the regulatory and management framework ("Management, Monitoring, Policy and Law") as well as to the design of structures and equipment ("Civil and Structural Engineering", "General Materials Science"), measurement and control devices ("Electrical and Electronic engineering") and energy-related issues ("Energy Engineering and Power Technology").

6.3 Niche Subcategories with the Highest Impact

Table 6 provides an additional level of analysis, particularly about the niches in which researchers are particularly interested. It shows the 20 Subcategories with the greatest impact (FWCI), contributing between 1.2 and 4.5% to the corpus. This list is dominated by "Agricultural and Biological Sciences" and "Earth and Planetary Sciences", with 4 Subcategories each. This is followed by "Engineering" with 3 Subcategories (Soleimani et al. 2023; Weiss et al. 2020; Jung et al. 2020).

Table 6 Main subcategories with the highest impact

Subject area	Subcategory	% of corpus	FWCI
Earth and planetary sciences	**"Geology"**	**2.2**	**3.4**
Earth and planetary sciences	**"Computers in Earth Sciences"**	**2.7**	**3.02**
Physics and astronomy	**"Condensed Matter Physics"**	**3.1**	**2.98**
Physics and astronomy	**"General Physics and Astronomy"**	**1.7**	**2.98**
Agricultural and biological sciences	**"Soil Science"**	**3.0**	**2.91**
Environmental science	**"Ecology"**	**2.3**	**2.91**
Chemistry	**"Physical and Theoretical Chemistry"**	**2.3**	**2.9**
Medicine	**"Public Health, Environ. and Occupational Health"**	**1.6**	**2.89**
Earth and planetary sciences	**"Oceanography"**	**1.4**	**2.85**
Agricultural and biological sciences	**"Ecology, Evolution, Behavior and Systematics"**	**1.8**	**2.84**
Agricultural and biological sciences	**"Agronomy and Crop Science"**	**2.9**	**2.83**
Engineering	**"Building and Construction"**	**3.3**	**2.79**
Agricultural and biological sciences	**"Forestry"**	**1.3**	**2.7**
Engineering	**"Engineering (miscellaneous)"**	**1.5**	**2.63**
Engineering	**"Safety, Risk, Reliability and Quality"**	**2.8**	**2.51**
Environmental science	**"Health, Toxicology and Mutagenesis"**	**3.0**	**2.44**
Materials science	**"Electronic, Optical and Magnetic Materials"**	**1.7**	**2.39**
Mathematics	**"Modeling and Simulation"**	**2.4**	**2.38**
Earth and planetary sciences	**"Earth-Surface Processes"**	**2.3**	**2.37**
Mathematics	**"Theoretical Computer Science"**	**1.2**	**2.37**

This list shows a tendency to develop the fundamental sciences related to the earth, water and matter ("Geology", "Condensed Matter Physics", "General Physics and Astronomy", "Soil Science", "Physical and Theoretical Chemistry", "Earth-Surface Processes") and Mathematics/Informatics ("Computers in Earth Sciences", "Modeling and Simulation", "Theoretical Computer Science").

The obvious intention is to observe and understand phenomena related to ("Ecology", "Oceanography", "Ecology, Evolution, Behavior and Systematics", "Agronomy and Crop Science", "Forestry", "Health, Toxicology and Mutagenesis").

At first sight, this knowledge is geared towards reinvesting and developing aspects linked to the tools and materials, both for measurement purposes and for design and implementation ("Electronic, Optical and Magnetic Materials" and "Building and Construction"). Finally, there is a very perceptible interest in issues relating to health and ecology ("Public Health, Environmental and Occupational Health", "Ecology, Evolution, Behavior and Systematics", "Safety, Risk, Reliability and Quality", "Health, Toxicology and Mutagenesis").

To conclude this overview, we noted that the theme studied was in a stage of accelerated development and that international collaboration was structuring this field. We have also established that interest is clearly focused on fundamental, and probably capital-intensive, questions, due to the resources required for observation and data acquisition activities for planetary phenomena.

Conversely, this could explain the importance of international collaboration, involving the major scientific nations, as well as the lack of interest shown by the corporate sector. Indeed, the industrial world does not seem to be convinced of the applicative and financial opportunities in this field. This finding prompts us to explore this area in greater depth, in an attempt to understand the nature of industrial collaboration in this field.

6.4 Relative Position of the Corporate Collaboration

The international collaboration corpus represents 29.8% of the total, with an FWCI of 2.26. Whereas the 1033 publications in industrial collaboration (3.5% of the total corpus) show a modest performance in terms of FWCI (1.59). In addition, only 0.8% of publications (239) received patent citations (376 citations). Despite China's central role in the overall corpus, the majority of these citing patents (341) are registered with the US Office (167), followed by the World Office (121) and the European Office (41). The 239 publications cited by patents also show a significantly higher FWCI (5.12). While only 18 of these 239 were published in industrial collaboration, with an FWCI of 4.46.

Figure 10 shows the performance of the various subsets of the overall corpus, in terms of the percentage of publications belonging to the world's top 10% (weighted) in terms of citations.

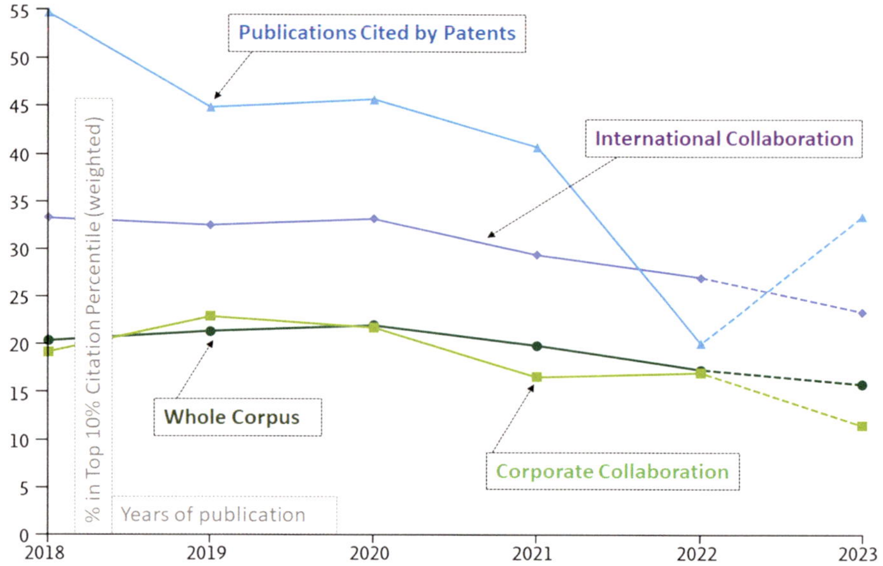

Fig. 10 Performance of the subsets of the general corpus

6.5 *Relationship Between Industry and Patents?*

Leaving aside the small proportion of the different sub-corpora in the total, Fig. 10 shows the poor performance of industrial collaboration, compared with the general corpus and so with that achieved through international collaboration. Conversely, it also shows the quality of international collaboration and the publications having an interest in the inventive activity (cited by patents).

Table 7 supports this observation. It shows that 36% of these 239 publications (cited by patents) are in international collaboration, with 5.61 FWCI. Nearly as many of these publications were produced within a single institution, with 3.27 FWCI. However, 28.5% of these 239 were published exclusively in national collaboration, with an FWCI of 6.79, even higher than that for international collaboration.

The results of this analysis are remarkable. In summary, they suggest that the majority of industrial collaborations (~98.3%) are not of immediate interest for the inventive activity. Conversely, the bulk of the inventive consistency in this theme,

Table 7 Performance of domestic and international collaboration

Collaboration type	%	Pub.	Cit./Pub.	FWCI
International collaboration	**36**	**86**	**98**	**5.61**
Only national. collab	**28.5**	**68**	**108.4**	**6.79**
Only institutional collab	**34.3**	**82**	**48.1**	**3.27**
No collaboration	**1.3**	**3**	**48.7**	**3.48**

even if modest, is likely to come from the academic world. Furthermore, 64% of the publications cited by patents are produced at the national level, with exceptional performance (6.79 FWCI) when this is done between local institutions (28.5% of cases).

One likely interpretation is that the importance of the subject and its future challenges explain the interest shown in it by the academic world (Elbasi et al. 2023; Siddiqui et al. 2022; Ray et al. 2023; Batisha 2023). Its precociousness and the weight of its prerequisites could justify the concentration of this research within a restricted group of leading scientific countries (Sudhakar 2023).

The same reasons may be behind the delay in the corporate world. In addition to the geographical and climatic specificities of each country, the geopolitical challenges of this theme may also suggest another point of view (Lopez-Gunn and Llamas 2008; Selby 2005; Helsenki Security Forum 2023).

6.6 *The Main Issues for Industrial Collaboration*

The previous results suggest that we should ask a question about the issues of interest for industrial collaboration in this field. Table 8 represents a selection of topics from the industrial collaboration sub-corpus. It is made up of two groups:

- The most prolific topics (+10 publications) and
- Those that contribute most to the topic (>1.3%).

While remaining within the structuring orientation of the general corpus (Sapitang et al. 2020; Girihagama et al. 2022; Li et al. 2019b; Linaza et al. 2021; Hou 2023), this list reveals a clear orientation towards topics essentially linked to the exploitation of some natural resources, in particular Oil, Gas and Petrol (Djanuar 2023; Zanganeh Kamali et al. 2022; Teixeira and Secchi 2019). There are also several issues relating to "wave energy"(Meng et al. 2022), "nuclear power" (Zhang et al. 2021) and the "paper industry" (Shan et al. 2022). Other cross-cutting industrial themes are also present, such as the mechanics of fluids, hydrogeological exploration, treatment of industrial effluents, etc.

These industrial issues generally have a scientific development model centred on other more endogenous themes than the application of AI-ML to the subject of water. This would probably explain the industrial "timidity" observed in this corpus. Conversely, this hypothesis also suggests a certain level of interference between these different themes and the theme under study (AI and Water) and that this could be interpreted as an opportunity to watch.

Furthermore, 61.2% of industrial collaborations are published at the national level (FWCI of 1.08). The remaining 38.8% are international collaborations (FWCI of 2.38). Admittedly, this ratio to international collaboration is significantly higher than the average for the general corpus (proportion of 29.8% and FWCI of 2.26), but this

Table 8 Main topics of the industrial collaboration

Topics	Pub.	% of topic	FWCI
"Object Detection; Deep Learning; IOU"	38	0.03	1.57
"Prediction; Flood Forecasting; Water Tables"	27	0.45	2.25
"Oil And Gas Fields; Bottom Hole Pressure; Reservoir Management"	16	1.93	1.01
"MERIS; Water Quality; Suspended Matter"	15	0.7	3.67
"Radar Altimetry; Flood Hazards; Remote Sensing"	14	0.31	2.7
"TBM; Tunnel; Hard Rock"	13	0.72	2.95
"Oil Well Flooding; Reservoir Simulation; Injectors"	12	5.24	1.8
"Petroleum Reservoirs; Ensemble Kalman Filter; Reservoir Simulation"	12	0.89	1.53
"Oil And Gas Fields; Petroleum Reservoir Evaluation; Rock Property"	11	1.64	1.19
"Shales; Silurian; Porosity"	10	0.13	1.85
"Artificial Neural Network; Wave Energy; Data Buoy"	8	1.72	3
"Nuclear Power Plants; Loss of Coolant Accidents; Accident"	8	1.45	0.68
"Oil Fields; Automatic Testing; Oil Well Flooding"	3	4.55	1.44
"Gas Lifts; Risers; Gas"	3	1.3	1.33
"Chokes; Flow Rate; Wellheads"	3	1.6	0.9
"Association Measure; T-Conorm; Artificial Intelligence"	3	2.75	0.46
"Water Treatment Plants; Artificial Neural Network; Raw Water"	2	2.41	2.35
"Carbon Dioxide; Equations of State; Brines"	2	1.32	1.9
"Activity Coefficients; Solvation; Liquid"	2	1.5	1.1
"Characteristic Curve; Oil Fields; Water"	2	1.92	0.93
"Pulp; Waste Liquors; Washing"	2	2.74	0.1

data also shows that the interest, usually expected for industrial applications, seems very weak for this corpus (Bikard et al. 2019).

This analysis reinforces the previous interpretation, which argues for an economic and industrial prematurity, accompanied by a geopolitical challenge for those nations that have the will and means to support the scientific and technical development of this field. Furthermore, the application of advances in AI-ML to the water field seems to bring a new dynamic to the various areas where knowledge about water can provide new scientific insights and ideas for development.

7 Brief Cross-Check Analysis on Incites

At this stage of the analysis, we have carried out a broad exploration of the bibliometric data from Scopus. Firstly, a general view of the evolution and distribution by country and major Subject Areas. Then, using the Scival platform, we took a closer look at the themes, collaborations and priority research questions. Finally, in the process of successive analyses, certain structuring trends emerged concerning the thematic dynamism, its maturity in terms of industrial potential, and the major research orientations.

7.1 Complementarity of the Two Frameworks

We are now carrying out an analytical cross-check using data from WOS, a comparable bibliographic database provided by Clarivate. This will enable us to look at this theme from a different perspective, due in particular to the differences between the two databases in terms of coverage, scientific nomenclature and analytical refinement (Singh et al. 2021). The aim is to compare the two perspectives to assess the similarity and relevance of the results obtained from the previous analysis on Scopus.

We are adopting the same earlier approach to construct and refine an equivalent corpus (Corp3). We thus identify the set of publications at the WOS level. The data is then transferred to Incites, Clarivate's analytical platform. Incites offers analytical features for WOS data, such as Scival for Scopus data.

Assuming the differences between Scopus and WOS, in terms of coverage, volume and scientific nomenclature, it is worth highlighting that the corpus (Corp3) identified at the WOS level shows the same major trends as those already observed at the Scopus level. In fact, in addition to the allure and dynamism of this corpus, we have established, for example, that the first 6 major domains, the most prolific, cover almost 87% of the overall corpus. These domains are also consistent with those of Scopus. This reinforces the preliminary analysis on Scopus and enables us to enrich our knowledge of this theme by exploring other dimensions using Incites (Bornmann and Leydesdorff 2013).

7.2 Main Research Topics

To detect the most important research issues in this corpus, we will analyse its performance in terms of "Citation Topics". This concept is similar to "Prominent Topics" in Scival. They consist of disjoint clusters of publications, representing single topics. They are built using a hierarchical classification algorithm, based on the density of the citation network between publications (Traag et al. 2019). Three hierarchical levels are then established: Macro (10 clusters), Meso (326) and Micro

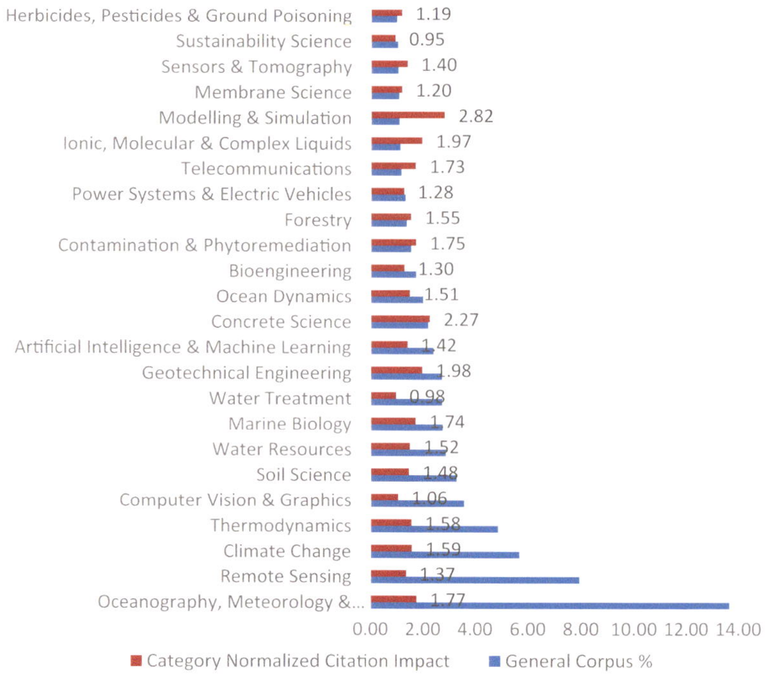

Fig. 11 Main meso topics of the general corpus in Incites

(2488) (Traag et al. 2019; Mejia et al. 2021). Clusters at the Macro and Meso levels are humanly labelled, while those at the Micro level are named, using an algorithm, based on keyword frequency. This gives greater relevance and facilitates interpretation when analysing clusters at the first two levels.

Thus, of Incites' 326 Meso Topics, 289 are represented in the corpus studied (Corp3). However, only 24 topics contribute more than 1% to the corpus (Fig. 11). These first 24 topics represent nearly 70% of the corpus studied. They also account for 76% of international collaboration and 68% of industrial collaboration (co-authored with researchers from corporate).

Finally, this list shows the same concentration profile of the main subjects as in Scopus. We can see also that the top topics on Incites are very coherent with those identified for the equivalent corpus on Scival. This establishes and highlights a structuring trend in this field of research.

Indeed, apart from the modest performance of the topics "Sustainability" and "Water Treatment", with a CNCI[3] (Understanding the metrics—incites benchmarking analytics—libguides at clarivate analytics 2024) of 0.95 and 0.98, the rest of the topics show high values of that index. Moreover, considering their high volume of publications, the first 4 most prolific topics (Fig. 11) are at outstanding CNCI levels

[3] CNCI "Category Normalised Citation Impact" is a metric of impact in WOS, broadly similar to FWCI in Scopus.

(Moustafa et al. 2022; Hu et al. 2019; Yuan 2020; Bonafilia et al. 2020; Yaseen et al. 2019). In addition, the 4 topics, "Geotechnical Engineering" (Panahi et al. 2020), "Concrete Science" (Zhao et al. 2022), "Ionic, Molecular and Complex Liquids" (Cheng et al. 2019; Andrade et al. 2020) and "Modelling and Simulation" (Almajid and Abu-Al-Saud 2022), achieve exceptional CNCI values, between ~2 and 3. All these facts reaffirm the broad interest in this theme and its dynamism, which is reflected in the CNCI scores of the first 24 topics.

We will use Incites to explore this cross-referencing in greater depth by looking at two complementary aspects: —the major themes of international collaboration and—the industrial and innovation potential.

7.3 *Effect of International Collaboration*

Except for "Geotechnical Engineering" and "Concrete Science", international collaboration does not exceed 40% at the level of each sub-corpus, particularly for the prolific disciplines (size of the sphere) that structure this research field (Fig. 12).

However, independently of the topic's size, this international openness seems to play a positive role in terms of citation performance (CNCI). This observation reinforces the elements established on the Scopus data (Fig. 10) and sheds light on the topics that benefit from this positive influence of international collaboration.

A few exceptions, such as "Modelling and Simulation" and "Ionic, Molecular and Complex Liquids", seem to be influenced through the conjunction with other parameters, probably in interference with that of the international collaboration.

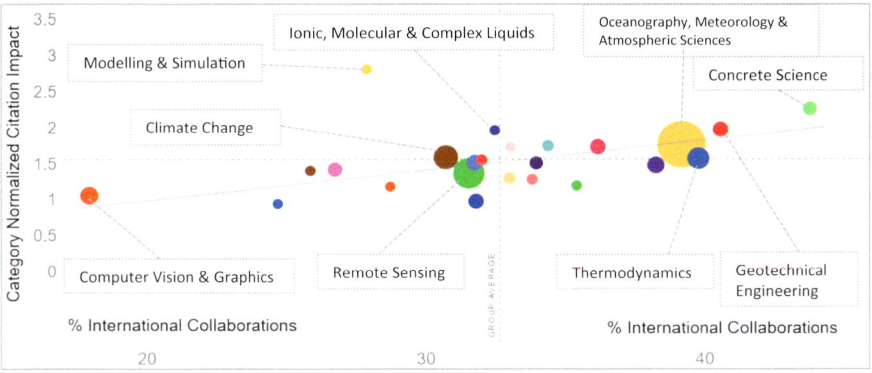

Fig. 12 Impact of international collaboration of the main topics

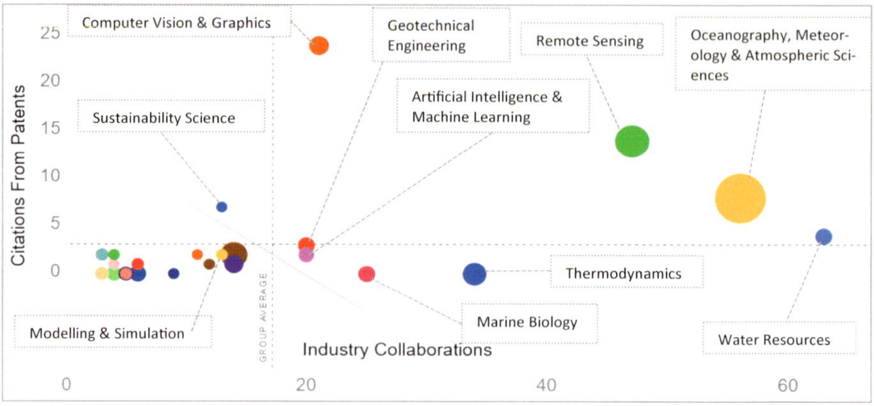

Fig. 13 Main meso topics related to industry and patenting activity

7.4 Industrial and Inventive Potential

All 24 topics are involved in industrial collaborations. However, only 8 of them collaborate on more than 20 publications with companies, including three of the four most prolific. Figure 13 also highlights the case of "Water Resources", which produces more than 8% of its publications with industry (63 publications). "Computer Vision and Graphics" (Li et al. 2020a, b; Zhou et al. 2022) stands out for its 24 patent citations, 10 more than "Remote Sensing", in second position. The performance of "Computer Vision and Graphics" is highlighted by its relatively small size and its modest industrial collaboration (21 pub.).

In short, Fig. 13 shows a reduced set of topics involved in industrial collaboration and even fewer, relevant to inventive activity (cited by patents). This also confirms the conclusions of the Scopus data analysis: prematurity balanced by a remarkable dynamism to be monitored.

8 Summary and Discussion

This study delves into the intersection of water research and AI-ML advances, providing a comprehensive analysis of the evolving landscape in this interdisciplinary field. The integration of AI-ML techniques into water research has led to a dynamic corpus of publications, with significant growth observed since 2008. This corpus is characterized by a strong emphasis on leveraging AI-ML to address key water research areas and aligns with the evolving AI models and practices.

Analysis of the corpus reveals a concentration of publications from the top 18 countries, highlighting their dominance either in publishing and international collaboration. This research shows a focus on engineering approaches, utilising technologies like "remote sensing", "optical engineering", and "photogrammetry" to observe environmental, hydrological, and geological phenomena. Data generated by these techniques are voluminous and require advanced computational methods for analysis, justifying the role of AI-ML in processing and interpreting such data.

Indeed, the most prolific research topics revolve around the detection and measurement of water-related phenomena, aiming to model and predict their behaviour for improved ecological management. Additionally, there is a growing interest in cross-disciplinary themes like "energy", "thermodynamics", "materials", and "chemistry". AI-ML techniques such as "artificial neural networks", "transfer learning", "long short-term memory", and "random forest" are prominent in the corpus and accentuate the issue of processing the vast amounts of data generated by water research. The study also highlights the main topics observed, like "water quality", "groundwater", "water levels", and "water resource management", with an emphasis on "forecasting" future trends.

Meanwhile, international collaboration is an important aspect, contributing to a consensual understanding of climate change issues and ecosystem dynamics. However, industrial collaboration in this area is currently limited, perhaps due to the complex nature of the field and its prerequisites, as well as countries' geopolitical and industrial interests.

In conclusion, this study provides valuable insights into the current trends and future directions of water research at the intersection with AI-ML. It highlights the potential to be developed in terms of industrial collaboration and innovative activity as well as scientific niches, which could constitute avenues for development and maturity.

Finally, consulting panels of experts could complete and enhance this analysis and guide future studies in this field. This can be an interesting perspective to go ahead with the exploration of this promising field.

References

Abdalla S et al (2021) Altimetry for the future: building on 25 years of progress. Adv Space Res 68(2):319–363. https://doi.org/10.1016/J.ASR.2021.01.022

Abdelmageed S, Tariq S, Boadu V, Zayed T (2022) Criteria-based critical review of artificial intelligence applications in water-leak management. Environ Rev 30(2):280–297. https://doi.org/10.1139/er-2021-0046

Adnan RM et al (2023) Application of novel binary optimized machine learning models for monthly streamflow prediction. Appl Water Sci 13(5). https://doi.org/10.1007/S13201-023-01913-6

Almajid MM, Abu-Al-Saud MO (2022) Prediction of porous media fluid flow using physics informed neural networks. J Pet Sci Eng 208. https://doi.org/10.1016/J.PETROL.2021.109205

Andrade MFC, Ko HY, Zhang L, Car R, Selloni A (2020) Free energy of proton transfer at the water-TiO2 interface from: ab initio deep potential molecular dynamics. Chem Sci 11(9):2335–2341. https://doi.org/10.1039/C9SC05116C

Azirar A, Boutracheh H (2019) La recherche scientifique et l'innovation dans le monde arabe. Rabat, July 2019. https://ires.ma/fr/publications/rapports-thematiques/la-recherche-scientifique-et-linnovation-dans-le-monde-arabe. Accessed 09 Apr 2023

Batisha A (2023) A lighthouse to future opportunities for sustainable water provided by intelligent water hackathons in the Arabsphere. Humanit Soc Sci Commun 10(1). https://doi.org/10.1057/s41599-023-02146-3

Batool D et al (2022) A hybrid approach to tea crop yield prediction using simulation models and machine learning. Plants 11(15). https://doi.org/10.3390/PLANTS11151925

Bawankule R, Gaikwad V, Kulkarni I, Kulkarni S, Jadhav A, Ranjan N (2023) Visual detection of waste using YOLOv8. In: International conference on sustainable computing and smart systems, ICSCSS 2023 - proceedings, pp 869–873.https://doi.org/10.1109/ICSCSS57650.2023.10169688

Bikard M, Vakili K, Teodoridis F (2019) When collaboration bridges institutions: the impact of university-industry collaboration on academic productivity. Organ Sci 30(2):426–445. https://doi.org/10.1287/orsc.2018.1235

Black J (1794) IV. An analysis of the waters of some hot springs in Iceland. Earth Environ Sci Trans R Soc Edinb 3(2):95–126. https://doi.org/10.1017/S0080456800020329

Blasch E, Majumder U, Zelnio E, Velten V (2020) Review of recent advances in AI/ML using the MSTAR data. In: Proceedings of SPIE - the international society for optical engineering. https://doi.org/10.1117/12.2559035

Bolibar J, Rabatel A, Gouttevin I, Zekollari H, Galiez C (2022) Nonlinear sensitivity of glacier mass balance to future climate change unveiled by deep learning. Nat Commun 13(1). https://doi.org/10.1038/S41467-022-28033-0

Bonafilia D, Tellman B, Anderson T, Issenberg E (2020) Sen1Floods11: a georeferenced dataset to train and test deep learning flood algorithms for sentinel-1. In: IEEE computer society conference on computer vision and pattern recognition workshops, vol 2020, pp 835–845, June 2020, https://doi.org/10.1109/CVPRW50498.2020.00113

Boretti A, Rosa L (2019) Reassessing the projections of the world water development report. NPJ Clean Water 2(1):1–6. https://doi.org/10.1038/s41545-019-0039-9

Bornmann L (2020) Bibliometrics-based decision trees (Bbdts) based on bibliometrics-based heuristics (bbhs): visualized guidelines for the use of bibliometrics in research evaluation. Quant Sci Stud 1(1):171–182. https://doi.org/10.1162/qss_a_00012

Bornmann L, Leydesdorff L (2013) The validation of (advanced) bibliometric indicators through peer assessments: a comparative study using data from InCites and F1000. J Informetr 7(2):286–291. https://doi.org/10.1016/J.JOI.2012.12.003

Boutracheh H, El Ansari R, Mejjad N, Moumen A (2023) Application of bibliometrics as a data mining technique for research prioritization: a case study of desalination research. In: Proceedings of the 6th international conference on networking, intelligent systems & security, pp 1–9, May 2023. https://doi.org/10.1145/3607720.3607779

Boutracheh H, El Bouhaddioui M, Moumen A (2024) Current research priorities on fog harvesting as a clean water resource: a bibliometric approach. In: E3S web of conferences, vol 489, p 05002, October 2024. https://doi.org/10.1051/E3SCONF/202448905002

Brown S (2015) Water imagery and the power and presence of God in the Gospel of John 72(3):289–298. https://doi.org/10.1177/0040573615601471

Brown T, Gutman SA (2019) Impact factor, eigenfactor, article influence, scopus SNIP, and SCImage journal rank of occupational therapy journals. Scand J Occup Ther 26(7):475–483. https://doi.org/10.1080/11038128.2018.1473489

Calver M, Bryant K, Wardell-Johnson G (2018) Quantifying the internationality and multidisciplinarity of authors and journals using ecological statistics. Scientometrics 115(2):731–748. https://doi.org/10.1007/s11192-018-2692-z

Chen Y, Fan R, Yang X, Wang J, Latif A (2018) Extraction of urban water bodies from high-resolution remote-sensing imagery using deep learning. Water (Switzerland) 10(5). https://doi.org/10.3390/W10050585

Chen L, Wu T, Wang Z, Lin X, Cai Y (2023) A novel hybrid BPNN model based on adaptive evolutionary Artificial Bee Colony Algorithm for water quality index prediction. Ecol Indic 146. https://doi.org/10.1016/J.ECOLIND.2023.109882

Cheng B, Engel EA, Behler J, Dellago C, Ceriotti M (2019) Ab initio thermodynamics of liquid and solid water. Proc Natl Acad Sci U S A 116(4):1110–1115. https://doi.org/10.1073/PNAS.1815117116

Cochran DS, Smith J, Mark BG, Rauch E (2022) Information model to advance explainable AI-based decision support systems in manufacturing system design, vol 525. LNNS. https://doi.org/10.1007/978-3-031-14317-5_5

Cucari N, Tutore I, Montera R, Profita S (2023) A bibliometric performance analysis of publication productivity in the corporate social responsibility field: outcomes of SciVal analytics. Corp Soc Responsib Environ Manag 30(1):1–16. https://doi.org/10.1002/csr.2346

de Solla Price D (1978) Cumulative advantage urn games explained: a reply to Kantor. J Am Soc Inf Sci 29(4):204–206. https://doi.org/10.1002/asi.4630290410

D'Amore G, Di Vaio A, Balsalobre-Lorente D, Boccia F (2022) Artificial intelligence in the water–energy–food model: a holistic approach towards sustainable development goals. Sustainability (Switzerland) 14(2). https://doi.org/10.3390/su14020867

Davis R, Panja P, McLennan J (2023) Integrated workflow for interpretation of satellite imageries using machine learning to assess and monitor algal blooms in Utah Lake, USA. Ecol Inform 75. https://doi.org/10.1016/J.ECOINF.2023.102033

Di Nunno F, Granata F (2023) Future trends of reference evapotranspiration in Sicily based on CORDEX data and Machine Learning algorithms. Agric Water Manag 280. https://doi.org/10.1016/J.AGWAT.2023.108232

Djanuar Y et al (2023) A novel approach to combine models to evaluate interwell connectivity in a waterflooded reservoir with limited injection history. In: Society of petroleum engineers - gas and oil technology showcase and conference, GOTS 2023. https://doi.org/10.2118/214268-MS

Donthu N, Kumar S, Mukherjee D, Pandey N, Lim WM (2021) How to conduct a bibliometric analysis: an overview and guidelines. J Bus Res 133:285–296. https://doi.org/10.1016/J.JBUSRES.2021.04.070

Dresbeck R (2015) SciVal. J Med Libr Assoc 103(3):164–166. https://doi.org/10.3163/1536-5050.103.3.018

Duguet T et al (2024) General practice-related MeSH terms in main journals: a bibliometric analysis from 2011 to 2021. Br J Gen Pract 74(739):e120–e125. https://doi.org/10.3399/BJGP.2023.0060

Elbasi E et al (2023) Artificial intelligence technology in the agricultural sector: a systematic literature review. IEEE Access 11:171–202. https://doi.org/10.1109/ACCESS.2022.3232485

Elsevier (2023) Topic prominence in SciVal. https://service.elsevier.com/app/answers/detail/a_id/28428/supporthub/scival/related/1/. Accessed 19 Nov 2023

Elsevier (2024) Fingerprint Engine|Powers Research Intelligence for institutions & funding bodies|Elsevier. https://www.elsevier.com/products/elsevier-fingerprint-engine. Accessed 06 Feb 2024

Elsheikh AH, El-Said EMS, Abd Elaziz M, Fujii M, El-Tahan HR (2023) Water distillation tower: experimental investigation, economic assessment, and performance prediction using optimized machine-learning model. J Clean Prod 388. https://doi.org/10.1016/J.JCLEPRO.2023.135896

Fang X et al. (2022) Deep reinforcement learning optimal control strategy for temperature setpoint real-time reset in multi-zone building HVAC system. Appl Therm Eng 212. https://doi.org/10.1016/J.APPLTHERMALENG.2022.118552

Fleming T VIII (1788) Account of a remarkable agitation of the waters of Loch Tay. Earth Environ Sci Trans R Soc Edinb 1(2):200–202. https://doi.org/10.1017/S0080456800029203

Gaagai A et al (2023) Application of water quality indices, machine learning approaches, and GIS to identify groundwater quality for irrigation purposes: a case study of Sahara Aquifer, Doucen Plain, Algeria. Water (Switzerland) 15(2). https://doi.org/10.3390/W15020289

Gagné T, Lapalme J, McQueen DV (2018) Multidisciplinarity in health promotion: a bibliometric analysis of current research. Health Promot Int 33(4):610–621. https://doi.org/10.1093/HEAPRO/DAX002
Gigerenzer G, Todd PM and ABC Research Group (2000) Simple heuristics that make us smart. In: Behavioral and brain sciences, vol 23. Cambridge University Press, pp 727–780
Ghosh H, Tusher MA, Rahat IS, Khasim S, Mohanty SN (2023) Water Quality Assessment Through Predictive Machine Learning. Lecture Notes in Networks and Systems, vol. 699. LNNS, pp 77–88. https://doi.org/10.1007/978-981-99-3177-4_6
Girihagama L et al (2022) Streamflow modelling and forecasting for Canadian watersheds using LSTM networks with attention mechanism. Neural Comput Appl 34(22):19995–20015. https://doi.org/10.1007/S00521-022-07523-8
Giwa SO, Sharifpur M, Goodarzi M, Alsulami H, Meyer JP (2021) Influence of base fluid, temperature, and concentration on the thermophysical properties of hybrid nanofluids of alumina–ferrofluid: experimental data, modeling through enhanced ANN, ANFIS, and curve fitting. J Therm Anal Calorim 143(6):4149–4167. https://doi.org/10.1007/S10973-020-09372-W
Guru Prasad MS, Agarwal J, Christa S, Aditya Pai H, Kumar MA, Kukreti A (2023) An improved water body segmentation from satellite images using MSAA-Net. In: 2023 international conference on machine intelligence for GeoAnalytics and remote sensing, MIGARS 2023. https://doi.org/10.1109/MIGARS57353.2023.10064508
Helsenki Security Forum 2023, New technologies reshape the geopolitics of water and climate. https://helsinkisecurityforum.fi/news/new-technologies-reshape-the-geopolitics-of-water-and-climate/. Accessed 08 Feb 2024
Hill WF (1960) Learning theory and the acquisition of values. Psychol Rev 67(5):317–331. https://doi.org/10.1037/h0043097
Hood WW, Wilson CS (2001) The literature of bibliometrics, scientometrics, and informetrics. Kluwer Academic Publishers
Hou M et al (2023) Efficient predictor of pressurized water reactor safety parameters by topological information embedded convolutional neural network. Ann Nucl Energy 192. https://doi.org/10.1016/J.ANUCENE.2023.110004
Hu P, Tong J, Wang J, Yang Y, De Oliveira Turci L (2019) A hybrid model based on CNN and Bi-LSTM for urban water demand prediction. In: 2019 IEEE congress on evolutionary computation, CEC 2019 - proceedings, pp 1088–1094, June 2019. https://doi.org/10.1109/CEC.2019.8790060
Huang Y, Khan SM (2021) Advances in AI and machine learning for education research. https://doi.org/10.1007/978-3-030-74394-9_11
Hutton C, Robison J (1790) V. Experiments on the expansive force of freezing water, made by Major Edward Williams of the Royal Artillery, at Quebec in Canada, in the years 1784 and 1785. Earth Environ Sci Trans R Soc Edinb 2(2):23–28. https://doi.org/10.1017/S0263593300027243
Ikram RMA et al (2023) Advanced hybrid metaheuristic machine learning models application for reference crop evapotranspiration prediction. Agronomy 13(1). https://doi.org/10.3390/AGRONOMY13010098
Ismail W, Niknejad N, Bahari M, Hendradi R, Zaizi NJM, Zulkifli MZ (2023) Water treatment and artificial intelligence techniques: a systematic literature review research. Environ Sci Pollut Res 30(28):71794–71812. https://doi.org/10.1007/s11356-021-16471-0
Jiang T, Su X, Zhang G, Zhang T, Wu H (2023) Estimating propagation probability from meteorological to ecological droughts using a hybrid machine learning copula method. Hydrol Earth Syst Sci 27(2):559–576. https://doi.org/10.5194/HESS-27-559-2023
Jung M et al (2020) Scaling carbon fluxes from eddy covariance sites to globe: synthesis and evaluation of the FLUXCOM approach. Biogeosciences 17(5):1343–1365. https://doi.org/10.5194/BG-17-1343-2020
Kanti PK, Sharma P, Maiya MP, Sharma KV (2023) The stability and thermophysical properties of Al2O3-graphene oxide hybrid nanofluids for solar energy applications: application of robust autoregressive modern machine learning technique. Solar Energy Mater Solar Cells 253. https://doi.org/10.1016/J.SOLMAT.2023.112207

Kanzola A, Papaioannou K, Petrakis P (2024) Unlocking society's standings in artificial intelligence. Technol Forecast Soc Change 200. https://doi.org/10.1016/j.techfore.2023.123106

Kasiselvanathan M, Sekar G, Prasad J, Lakshminarayanan S, Sharanya C. An IoT based agricultural management approach using machine learning. In: International conference on innovative data communication technologies and application, ICIDCA 2023 - proceedings, pp 61–65. https://doi.org/10.1109/ICIDCA56705.2023.10099598

Katimbo A et al (2023) Evaluation of artificial intelligence algorithms with sensor data assimilation in estimating crop evapotranspiration and crop water stress index for irrigation water management. Smart Agric Technol 4. https://doi.org/10.1016/J.ATECH.2023.100176

Kersh BY (1958) The adequacy of 'meaning' as an explanation for the superiority of learning by independent discovery. J Educ Psychol 49(5):282–292. https://doi.org/10.1037/h0044500

Khosravi K, Rezaie F, Cooper JR, Kalantari Z, Abolfathi S, Hatamiafkoueieh J (2023) Soil water erosion susceptibility assessment using deep learning algorithms. J Hydrol (Amst) 618. https://doi.org/10.1016/J.JHYDROL.2023.129229

Kumari S, Chowdhry J, Garg MC (2024) AI-enhanced adsorption modeling: challenges, applications, and bibliographic analysis. J Environ Manag 351. https://doi.org/10.1016/j.jenvman.2023.119968

Kuzma S, Saccoccia L, Chertock M (2023) 25 countries face extremely high water stress. World Resources Institute. https://www.wri.org/insights/highest-water-stressed-countries. Accessed 28 Oct 2023

Larivière V, Sugimoto CR (2018) Mesurer la science. Les Presses de l'Université de Montréal

Li C et al (2020a) An underwater image enhancement benchmark dataset and beyond. IEEE Trans Image Process 29:4376–4389. https://doi.org/10.1109/TIP.2019.2955241

Li D, Chen D, Jin B, Shi L, Goh J, Ng SK (2019a) MAD-GAN: Multivariate anomaly detection for time series data with generative adversarial networks. Lecture notes in computer science (including subseries Lecture notes in artificial intelligence and Lecture notes in bioinformatics), vol 11730. LNCS, pp 703–716. https://doi.org/10.1007/978-3-030-30490-4_56

Li W, He C, Fang J, Zheng J, Fu H, Yu L (2019) Semantic segmentation-based building footprint extraction using very high-resolution satellite images and multi-source GIS data. Remote Sens (Basel) 11(4). https://doi.org/10.3390/RS11040403

Li C, Anwar S, Porikli F (2020) Underwater scene prior inspired deep underwater image and video enhancement. Pattern Recognit 98. https://doi.org/10.1016/J.PATCOG.2019.107038

Linaza MT et al (2021) Data-driven artificial intelligence applications for sustainable precision agriculture. Agronomy 11(6). https://doi.org/10.3390/AGRONOMY11061227

Liu R, Xiao Y, Ma Y, Cui T, An J (2022) Red tide detection based on high spatial resolution broad band optical satellite data. ISPRS J Photogramm Remote Sens 184:131–147. https://doi.org/10.1016/J.ISPRSJPRS.2021.12.009

Liu Z et al (2023b) Remote sensing and geostatistics in urban water-resource monitoring: a review. Mar Freshw Res 74(9–10):747–765. https://doi.org/10.1071/MF22167

Liu Z, Qiu Y, Jafarinejad S (2023a) Artificial intelligence application to the nexus of renewable energy, water, and the environment. https://doi.org/10.1016/B978-0-443-13439-5.00012-0

Lo YK, Fu B (2023) Water in the mencius: correlative reasoning, conceptual metaphor, and/or sacred performative narrative? Religions 14(6):710. https://doi.org/10.3390/REL14060710

Lopez-Gunn E, Llamas MR (2008) Re-thinking water scarcity: can science and technology solve the global water crisis? Nat Resour Forum

Martens HH (1959) Two notes on machine 'learning.' Inf Control 2(4):364–379. https://doi.org/10.1016/S0019-9958(59)80014-0

McCarroll M, Hamann H (2020) What we know about water: a water literacy review. Water 12(10):2803. https://doi.org/10.3390/W12102803

Meisch SP (2019) I want to tell you a story: how narrative water ethics contributes to re-theorizing water politics. Water 11(4):631. https://doi.org/10.3390/W11040631

Mejia C, Wu M, Zhang Y, Kajikawa Y (2021) Exploring topics in bibliometric research through citation networks and semantic analysis. Front Res Metr Anal 6. https://doi.org/10.3389/FRMA.2021.742311/FULL

Meller LLT et al (2023) Identifying core journals in otolaryngology: a bibliometric analysis. Laryngoscope 133(12):3346–3352. https://doi.org/10.1002/lary.30709

Meng F, Xu D, Song T (2022) ATDNNS: an adaptive time–frequency decomposition neural network-based system for tropical cyclone wave height real-time forecasting. Futur Gener Comput Syst 133:297–306. https://doi.org/10.1016/J.FUTURE.2022.03.029

Min K, Cuiffi JD, Mathers RT (2020) Ranking environmental degradation trends of plastic marine debris based on physical properties and molecular structure. Nat Commun 11(1). https://doi.org/10.1038/S41467-020-14538-Z

Minsky M (1961) Steps toward artificial intelligence. Proc IRE 49(1):8–30. https://doi.org/10.1109/JRPROC.1961.287775

Mohammadian E, Hadavimoghaddam F, Kheirollahi M, Jafari M, Chenlu X, Liu B. Probing Solubility and pH of CO2 in aqueous solutions: Implications for CO2 injection into oceans. J CO2 Utiliz 71. https://doi.org/10.1016/J.JCOU.2023.102463

Mokhnacheva YV, Tsvetkova VA (2021) Development of research topics based on the terminological approach (for example, immunology and microbiology according to scopus–SciVal data). Sci Tech Inf Process 48(2):139–145. https://doi.org/10.3103/S0147688221020106

Mongeon P, Paul-Hus A (2016) The journal coverage of web of science and scopus: a comparative analysis. Scientometrics 106(1):213–228. https://doi.org/10.1007/S11192-015-1765-5

Mostafa RR, Kisi O, Adnan RM, Sadeghifar T, Kuriqi A (2023) Modeling potential evapotranspiration by improved machine learning methods using limited climatic data. Water (Switzerland) 15(3). https://doi.org/10.3390/W15030486

Moustafa EB, Hammad AH, Elsheikh AH (2022) A new optimized artificial neural network model to predict thermal efficiency and water yield of tubular solar still. Case Stud Therm Eng 30. https://doi.org/10.1016/J.CSITE.2021.101750

Nasr M, Islam MM, Shehata S, Karray F, Quintana Y (2021) Smart healthcare in the age of AI: recent advances, challenges, and future prospects. IEEE Access 9:145248–145270. https://doi.org/10.1109/ACCESS.2021.3118960

Navya P, Sudha D (2023) Artificial intelligence-based robot for harvesting, pesticide spraying and maintaining water management system in agriculture using IoT. AIP Conf Proc 2523. https://doi.org/10.1063/5.0110258

Nazar S, Yang J, Amin MN, Khan K, Javed MF, Althoey F (2023) Formulation of estimation models for the compressive strength of concrete mixed with nanosilica and carbon nanotubes. Dev Built Environ 13. https://doi.org/10.1016/J.DIBE.2022.100113

Oestigaard T (2021) Water and religion. In: Oxford research encyclopedia of anthropology. https://doi.org/10.1093/ACREFORE/9780190854584.013.477

Pahlevan N et al (2020) Seamless retrievals of chlorophyll-a from Sentinel-2 (MSI) and Sentinel-3 (OLCI) in inland and coastal waters: a machine-learning approach. Remote Sens Environ 240. https://doi.org/10.1016/J.RSE.2019.111604

Panahi M, Sadhasivam N, Pourghasemi HR, Rezaie F, Lee S (2020) Spatial prediction of groundwater potential mapping based on convolutional neural network (CNN) and support vector regression (SVR). J Hydrol (Amst) 588. https://doi.org/10.1016/J.JHYDROL.2020.125033

Peng W, Karimi Sadaghiani O (2023) Enhancement of quality and quantity of woody biomass produced in forests using machine learning algorithms. Biomass and Bioenergy 175. https://doi.org/10.1016/J.BIOMBIOE.2023.106884

Perkmann M, Walsh K (2009) The two faces of collaboration: impacts of university-industry relations on public research. Ind Corp Chang 18(6):1033–1065. https://doi.org/10.1093/ICC/DTP015

Premack D (1959) Toward empirical behavior laws: I. Positive reinforcement. Psychol Rev 66(4):219–233. https://doi.org/10.1037/h0040891

Price DJDS (1965) The scientific foundations of science policy. Nature 206(4981):233–238. https://doi.org/10.1038/206233a0
Price DDS (1976) A general theory of bibliometric and other cumulative advantage processes. J Am Soc Inf Sci 27(5):292–306. https://doi.org/10.1002/asi.4630270505
Purkayastha A, Palmaro E, Falk-Krzesinski HJ, Baas J (2019) Comparison of two article-level, field-independent citation metrics: Field-Weighted Citation Impact (FWCI) and Relative Citation Ratio (RCR). J Informetr 13(2):635–642. https://doi.org/10.1016/J.JOI.2019.03.012
Pyo JC et al (2019) A convolutional neural network regression for quantifying cyanobacteria using hyperspectral imagery. Remote Sens Environ 233. https://doi.org/10.1016/J.RSE.2019.111350
Raissi M, Perdikaris P, Karniadakis GE (2019) Physics-informed neural networks: A deep learning framework for solving forward and inverse problems involving nonlinear partial differential equations. J Comput Phys 378:686–707. https://doi.org/10.1016/J.JCP.2018.10.045
Rawat P, Bajaj M, Sharma V, Vats S (2023) A comprehensive analysis of the effectiveness of machine learning algorithms for predicting water quality. In: 2023 international conference on innovative data communication technologies and application, ICIDCA 2023 - Proceedings, pp 1108–1114. https://doi.org/10.1109/ICIDCA56705.2023.10099968
Ray SS, Verma RK, Singh A, Ganesapillai M, Kwon YN (2023) A holistic review on how artificial intelligence has redefined water treatment and seawater desalination processes. Desalination 546. https://doi.org/10.1016/j.desal.2022.116221
Reer A, Wiebe A, Wang X, Rieger JW (2023) FAIR human neuroscientific data sharing to advance AI driven research and applications: legal frameworks and missing metadata standards. Front Genet 14. https://doi.org/10.3389/fgene.2023.1086802
Richter A, Ng KTW, Fallah B (2019) Bibliometric and text mining approaches to evaluate landfill design standards. Scientometrics 118(3):1027–1049. https://doi.org/10.1007/S11192-019-03011-4
Rostaing H (1996) La bibliometrie et ses techniques, Outils et méthodes. Éditions Sciences de la Société AND CRRM
Ruidas D, Pal SC, Towfiqul Islam ARM, Saha A (2023) Hydrogeochemical evaluation of groundwater aquifers and associated health hazard risk mapping using ensemble data driven model in a water scares plateau region of Eastern India. Expo Health 15(1):113–131. https://doi.org/10.1007/S12403-022-00480-6
Santoro E, Monin B (2023) The AI effect: people rate distinctively human attributes as more essential to being human after learning about artificial intelligence advances. J Exp Soc Psychol 107. https://doi.org/10.1016/j.jesp.2023.104464
Sapitang M, Ridwan WM, Kushiar KF, Ahmed AN, El-Shafie A (2020) Machine learning application in reservoir water level forecasting for sustainable hydropower generation strategy. Sustainability (Switzerland) 12(15). https://doi.org/10.3390/SU12156121
Selby J (2005) The geopolitics of water in the middle east: fantasies and realities. Third World Q 26(2):329–349. https://doi.org/10.1080/0143659042000339146
Shahmansouri AA, Yazdani M, Ghanbari S, Bengar HA, Jafari A, Farrokh Ghatte H (2021) Artificial neural network model to predict the compressive strength of eco-friendly geopolymer concrete incorporating silica fume and natural zeolite. J Clean Prod 279. https://doi.org/10.1016/J.JCLEPRO.2020.123697
Shan WJ, Liu B, Tang W (2022) Soft sensing and multi-objective optimization of pulp washing process based on data-driven technology. Palpu Chongi Gisul/j Korea Tech Assoc Pulp Paper Ind 54(4):57–74. https://doi.org/10.7584/JKTAPPI.2022.08.54.4.57
Shubik M (1960) Bibliography on simulation, gaming, artificial intelligence and allied topics. J Am Stat Assoc 55(292):736–751. https://doi.org/10.1080/01621459.1960.10483374
Siddiqui A, Siddiqui M, Kulkarni N (2022) Artificial intelligence in water conservation: a meta-analysis study. Prabandhan: Indian J Manag 15(2):24–41. https://doi.org/10.17010/pijom/2022/v15i3/160407
Siegel S, Goldstein DA (1959) Decision-making behavior in a two-choice uncertain outcome situation. J Exp Psychol 57(1):37–42. https://doi.org/10.1037/h0045959

Singh VK, Singh P, Karmakar M, Leta J, Mayr P (2021) The journal coverage of web of science, scopus and dimensions: a comparative analysis. Scientometrics 126(6):5113–5142. https://doi.org/10.1007/S11192-021-03948-5

Soleimani S, Heydari A, Fattahi M, Motamedisade A (2023) Calcium alginate hydrogels reinforced with cellulose nanocrystals for methylene blue adsorption: synthesis, characterization, and modelling. Ind Crops Prod 192. https://doi.org/10.1016/J.INDCROP.2022.115999

Sudhakar M (2023) Artificial intelligence applications in water treatment and water resource assessment: challenges, innovations, and future directions. https://doi.org/10.4018/979-8-3693-0044-2.ch014.

Tariq A, Qin S (2023) Spatio-temporal variation in surface water in Punjab, Pakistan from 1985 to 2020 using machine-learning methods with time-series remote sensing data and driving factors. Agric Water Manag 280. https://doi.org/10.1016/J.AGWAT.2023.108228

Teixeira AF, Secchi AR (2019) Machine learning models to support reservoir production optimization. IFAC-PapersOnLine 52(1):498–501. https://doi.org/10.1016/J.IFACOL.2019.06.111

Ting L, Baijun Z, Yongsheng Z, Shun Y (2021) Ship detection algorithm based on improved YOLO V5. In: Proceedings - 2021 6th international conference on automation, control and robotics engineering, CACRE 2021, pp 483–487.https://doi.org/10.1109/CACRE52464.2021.9501331

Traag VA, Waltman L, van Eck NJ (2019) From Louvain to Leiden: guaranteeing well-connected communities. Sci Rep 9(1):1–12. https://doi.org/10.1038/s41598-019-41695-z

Uddin MG, Nash S, Rahman A, Olbert AI (2023) Performance analysis of the water quality index model for predicting water state using machine learning techniques. Process Saf Environ Prot 169:808–828. https://doi.org/10.1016/J.PSEP.2022.11.073

Understanding the metrics - incites benchmarking & analytics - libguides at clarivate analytics. https://clarivate.libguides.com/incites_ba/understanding-indicators. Accessed 09 Feb 2024

UNESCO and World Water Assessment Programme (2023) The United Nations World Water Development Report 2023: partnerships and cooperation for water. https://unesdoc.unesco.org/ark:/48223/pf0000384655. Accessed 28 Oct 2023

van Eck NJ, Waltman L (2010) Software survey: VOSviewer, a computer program for bibliometric mapping. Scientometrics 84(2):523–538. https://doi.org/10.1007/S11192-009-0146-3/FIGURES/7

Vicki Wanatasanappan V, Kumar Kanti P, Sharma P, Husna N, Abdullah MZ (2023) Viscosity and rheological behavior of Al2O3-Fe2O3/water-EG based hybrid nanofluid: a new correlation based on mixture ratio. J Mol Liq 375. https://doi.org/10.1016/J.MOLLIQ.2023.121365

Walle H, De Runz C, Serres B, Venturini G (2022) A survey on recent advances in AI and vision-based methods for helping and guiding visually impaired people. Appl Sci (Switzerland) 12(5). https://doi.org/10.3390/app12052308

Wallin JA (2005) Bibliometric methods: pitfalls and possibilities. Basic Clin Pharmacol Toxicol 97(5):261–275. https://doi.org/10.1111/j.1742-7843.2005.pto_139.x

Waltman L, van Eck NJ, van Leeuwen TN, Visser MS (2013) Some modifications to the SNIP journal impact indicator. J Informetr 7(2):272–285. https://doi.org/10.1016/j.joi.2012.11.011

Wang S, Peng H, Hu Q, Jiang M (2022) Analysis of runoff generation driving factors based on hydrological model and interpretable machine learning method. J Hydrol Reg Stud 42. https://doi.org/10.1016/J.EJRH.2022.101139

Wei X, Zhang L, Yang HQ, Zhang L, Yao YP (2021) Machine learning for pore-water pressure time-series prediction: application of recurrent neural networks. Geosci Front 12(1):453–467. https://doi.org/10.1016/J.GSF.2020.04.011

Weiss M, Jacob F, Duveiller G (2020) Remote sensing for agricultural applications: a meta-review. Remote Sens Environ 236. https://doi.org/10.1016/J.RSE.2019.111402

Wen L, Lu Y, Li H, Long S, Li J (2020) Detecting of research front topic in artificial intelligence based on SciVal. In: ACM international conference proceeding series, pp 145–149. https://doi.org/10.1145/3421766.3421799

Wijaya J, Oh S (2023) Machine learning reveals the complex ecological interplay of microbiome in a full-scale membrane bioreactor wastewater treatment plant. Environ Res 222. https://doi.org/10.1016/J.ENVRES.2023.115366

Xiao C et al (2023) Prediction of soil salinity parameters using machine learning models in an arid region of northwest China. Comput Electron Agric 204. https://doi.org/10.1016/J.COMPAG.2022.107512

Xu J, Mahmood H, Xiao H, Anderlini E, Abusara M (2021) Electric water heaters management via reinforcement learning with time-delay in isolated microgrids. IEEE Access 9:132569–132579. https://doi.org/10.1109/ACCESS.2021.3112817

Yaseen ZM, Sulaiman SO, Deo RC, Chau KW (2019) An enhanced extreme learning machine model for river flow forecasting: state-of-the-art, practical applications in water resource engineering area and future research direction. J Hydrol (AMST) 569:387–408. https://doi.org/10.1016/J.JHYDROL.2018.11.069

Yuan Q et al (2020) Deep learning in environmental remote sensing: achievements and challenges. Remote Sens Environ 241. https://doi.org/10.1016/J.RSE.2020.111716

Zanganeh Kamali M et al (2022) Permeability prediction of heterogeneous carbonate gas condensate reservoirs applying group method of data handling. Mar Pet Geol 139. https://doi.org/10.1016/J.MARPETGEO.2022.105597

Zebakh S, Arvanitis R, Boutracheh H, Sadiki M (2017) Trends in the Moroccan agricultural research: an exploratory bibliometric analysis (2005–2015). Revue Marocaine des Sciences Agronomiques et Vétérinaires 5(3):255–268. www.agrimaroc.org/index.php/Actes_IAVH2/article/download/546/496/. Accessed 09 Apr 2023

Zhang W, Wu C, Tang L, Gu X, Wang L (2023) Efficient time-variant reliability analysis of Bazimen landslide in the Three Gorges Reservoir Area using XGBoost and LightGBM algorithms. Gondwana Res 123:41–53. https://doi.org/10.1016/J.GR.2022.10.004

Zhang L et al (2021) System-level anomaly detection for nuclear power plants using variational graphauto-encoders. In: Proceedings of 2021 IEEE international conference on sensing, diagnostics, prognostics, and control, SDPC 2021, pp 180–185. https://doi.org/10.1109/SDPC52933.2021.9563515

Zhao L, Dai T, Qiao Z, Sun P, Hao J, Yang Y (2020) Application of artificial intelligence to wastewater treatment: a bibliometric analysis and systematic review of technology, economy, management, and wastewater reuse. Process Saf Environ Prot 133:169–182. https://doi.org/10.1016/J.PSEP.2019.11.014

Zhao Y, Hu H, Song C, Wang Z (2022) Predicting compressive strength of manufactured-sand concrete using conventional and metaheuristic-tuned artificial neural network. Measurement (Lond) 194. https://doi.org/10.1016/J.MEASUREMENT.2022.110993

Zhou Z, Zhang J, Gong C (2022) Automatic detection method of tunnel lining multi-defects via an enhanced You Only Look Once network. Comput-Aided Civ Infrastruct Eng 37(6):762–780. https://doi.org/10.1111/MICE.12836

A Bibliometric Analysis and Classification of Research on Water Resources Management Based on 17SDGs and ANZSRC Indicators

Nezha Mejjad, Aniss Moumen, Hicham Boutracheh, Ismail Hilal, Mohamed Qurtobi, and Mohamed El Bouhaddioui

Abstract Water scarcity has become a major global concern due to climate change, excessive use and pollution. Establishing and developing strategies and approaches to manage available water resources is challenging under the current natural and anthropogenic pressures. The present study analyzes the current research trend on Water Resource Management (WRM) by conducting a bibliometric analysis using Vosviewer software (version 1.6.20). The study also analyzes the research categories based on the 17 Sustainable Development Goals (17SDGs) and the Australian and New Zealand Standard Research Classification (ANZSRC). Based on the two classifications, the sustainable and effective management of water resources at the scientific level requires the integration of interdisciplinary research projects, while at the societal, economic, and strategic levels, the 17 SDGs must all be taken into account when establishing WRM projects. The results of the bibliometric analysis show that the current focus is on climate variability and its impact on food security. China is found to be the hotspot country with the highest number of publications, followed by the United States, while a lack of publications on WRM was found in developing countries that reported suffering from water scarcity and climate impacts. These findings are crucial for future research on WRM as they provide the basic knowledge about research trends in this research field.

N. Mejjad (✉) · I. Hilal · M. Qurtobi
National Centre for Nuclear Energy, Science and Technology, 14000 Kénitra, Morocco
e-mail: mejjadnezha@gmail.com

A. Moumen · H. Boutracheh
Ibn Tofail University, National School of Applied Sciences, 14000 Kénitra, Morocco

H. Boutracheh
CNRST, IMIST, National Center for Scientific and Technical Research, Rabat, Morocco

M. El Bouhaddioui
National School of Mines, B.P. 753, Agdal, 14000 Rabat, Morocco

N. Rebai et al. (eds.), *GIS, Applied Computing and Data Science for Water Management*, Lecture Notes in Geoinformation and Cartography,
https://doi.org/10.1007/978-3-031-63038-5_2

1 Introduction

The world is facing a water crisis due to climate change, chemical pollution, overexploitation of water resources, mismanagement, and poor governance.

The management of water resources has become challenging as its exploitation is uncontrolled and human activities continue to degrade its quality (Huo et al. 2008; Bruneau, 2009; Xia et al. 2012; Dey and Mishra 2017; Chen et al. 2020; Mohammadi Ghaleni and Ebrahimi 2015). This has led to water scarcity in many countries around the world. According to UNICEF, nearly two-thirds of the world's population faces severe water scarcity for about one month per year (UNICEF 2023).

The proper management of available water resources is an urgent necessity. Water scarcity threatens human existence by affecting food availability and security, threatening socio-economic sectors, and endangering human well- being (Rosegrant et al. 2009; Ripoll et al. 2014; Mancosu et al. 2015; Mohamed 2019; Karan & Asgari 2021; Mejjad et al. 2021). Numerous studies have shown that human activities contribute directly (excessive use of water for industrial and agricultural purposes) and indirectly (chemical and organic contamination) to the scarcity and degradation of water resources.

Protecting and conserving available water resources is crucial. Therefore, analyzing and assessing water contamination levels can contribute to the development of strategies and approaches to address water pollution issues and prevent water losses. Additionally, pollution assessments can identify potential sources of pollutants, allowing for the remediation of polluted water.

The success of water management strategies and approaches depends on considering all stakeholders, including governments, scientists, researchers, civil society, and water resource consumers such as farmers. Gleick and Heberger (2014) noted that almost 70% of global freshwater withdrawals are used for agriculture. Water use rates vary across the world, both geographically and based on income level (Doungmanee 2016; Ritchie and Roser 2023). In developing countries, agriculture accounts for the highest percentage, ranging from 70 to 90%, while in developed countries it is around 41% (Ritchie and Roser 2023). Therefore, WRM strategies should prioritize sectors with the highest water use rates.

The purpose of this chapter is to examine trends in research related to the assessment, management, and governance of water resources. The study also analyzes research on WRM based on the 17 sustainable development goals (17SDGs) and the Australian and New Zealand Standard Research Classification (ANZSRC) classifications. A bibliometric analysis was also conducted using the Vosviewer software tool to examine scientific publications on WRM. The purpose is to analyze research trends in the field of water pollution and management and identify the main gaps that need to be addressed.

2 Methods and Materials

2.1 Data Collection

Data search was carried out in Dimensions database, where the used keywords were as follow "Water AND resource AND management", "Water AND resource AND pollution". The search was based on abstracts and titles, resulting in 85,555 documents for the 'Water AND resource AND management' query, including full papers, short papers (proceedings), chapters, and preprints. The downloaded file contains the 500 most relevant documents to our search query. For the second search query, 'Water AND resource AND pollution', 25,277 publications were found. The search was conducted only in English and included policy and technical reports (Fig. 1).

2.2 Bibliometric Analysis: Vosviewer

Bibliometric study is a valuable tool that allows better understanding of the current knowledge on water resources management and governance. Such analysis would help in developing new research projects to fill gaps and address key issues and challenges related to water resources management. In the present study, we used Vosviewer software (version 1.6.20) to create a network map of keywords related to our search query to assess the distribution of publications among countries.

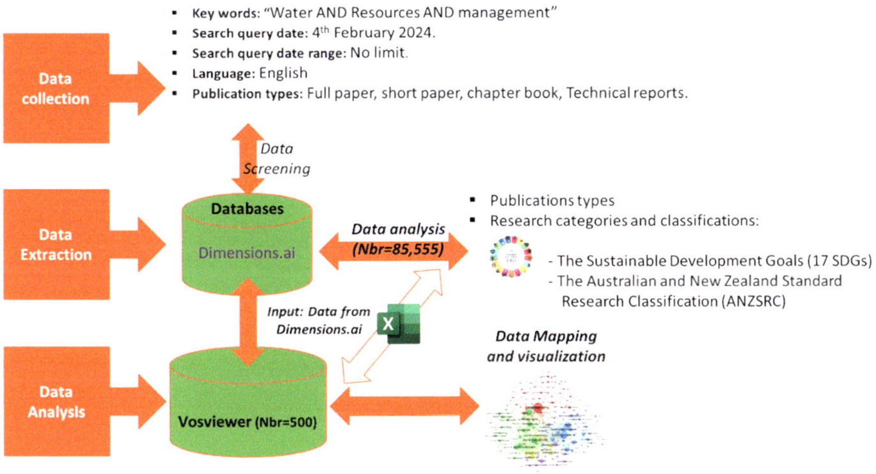

Fig. 1 Study's flowchart

3 Results

3.1 Chronology of Publications

3.1.1 Type of Publications

The analysis of publication types according to each search query may indicate that water resources research links water quality with management strategies. More than 70% of the publications published between 1900 and 2024 were articles, followed by chapters (15–17%) and proceedings (Fig. 2). 4% of the total publications found in the Dimensions database related to the water resources management search query were published in Journal of Water (2009 documents, 2%) and Journal of Hydrology (1407 documents, 2%), followed by Journal of the Science of the Total Environment, Water Resources Management, and Sustainability (1407, 1129, 1004 documents, respectively 1%).

3.1.2 Categories of Publications

The analysis of publication categories reflects the current interest of researchers in the field of water pollution and management. The publication categories are classified according to the Fields of Research (ANZSRC 2020) and the 17 Sustainable Development Goals. The Australian and New Zealand Standard Research Classification (ANZSRC) is used to measure and analyze research and development in all fields of research and education in Australia and New Zealand. It consists of three classifications, one of which is represented in the dimensions database as Fields of Research (FoR). The treemap of research category based on research field (ANZSRC) shows that the search query "water management resources" related publication research are

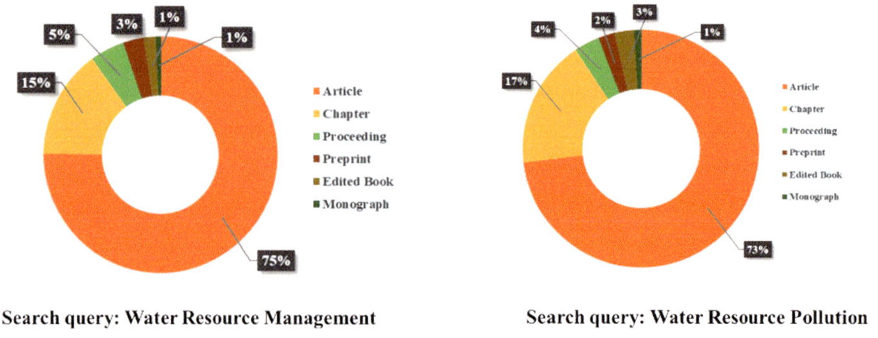

Fig. 2 Publications types according to each search query (Data retrieved from Dimension.ai, (04 February 2024)

mainly on earth sciences (33,904 documents), environmental sciences (23,486 documents), hydrology (21,049 documents), engineering (15,504 documents), environmental management (14,113 documents), agricultural, veterinary and food sciences (8920 documents), and economics (8,125 documents) (Fig. 3). Thus, WRM spans multiple fields of research, which means that addressing water management challenges requires a multidisciplinary assessment. Publications related to WRM are also published in other fields such as atmospheric sciences, human society, biological sciences, law and legal studies, and civil engineering with a range of documents between 7940 and 4027. It should be noted that proper and safe management of water resources can take several forms and involves the knowledge and expertise to work at the intersection of geology, mathematics, geography, meteorology, biology, political science, social science and even psychology (Cech 2018).

According to the 17 Sustainable Development Goals (SDGs), the publications related to the search query are mainly linked to the sixth goal, water and sanitation, which aims to ensure clean drinking water and sanitation for all and to ensure the sustainability of water resources management. 15,097 publications were linked to the SDG on life on land. The 15th SDG aims to conserve, restore and promote the sustainable use and exploitation of terrestrial ecosystem resources, which include water resources. Life on land depends on the quality and availability of water resources, which means water management is critical to achieving this goal. Climate change goal includes 9605 publications, as it severely affects the availability of water resources and caused water scarcity in many countries around the world (Sukanya and Joseph 2023).

WRM-related publications are also categorized within the other 14 SDGs, suggesting that WRM is directly and indirectly related to these goals:

- ***SDG2 Zero Hunger*** (6383 documents): Aims to ensure access to food for all. The quality and availability of water resources can affect food production, preparation and quality. Therefore, sustainable water management is critical to achieving food security (Taka et al. 2021).
- ***SDG 7 Affordable and clean energy*** (4339 documents): Aims to ensure access to clean and affordable energy, which will help reduce dependence on non-renewable energy that contributes to climate change and affects water availability and quality. In addition, the lack of electricity is an obstacle to water treatment, making access to clean and safe drinking water a challenge (Wang et al. 2019). In this sense, the success of water resource management also depends on the availability of clean and affordable energy.
- ***SDG 14 Life under water*** (3951 documents): Life under water depends mainly on water quality. Fish are a source of food, so if water quality deteriorates, it can affect food security and the blue economy in coastal countries. Mismanagement of water resources and water pollution can affect groundwater and transport pollutants to coastal and marine environments (Mejjad et al. 2021), which can threaten life under water.
- ***SDG 11 Sustainable cities and communities*** (2514 documents): ensuring sustainable cities and communities depends on ensuring sustainable management of

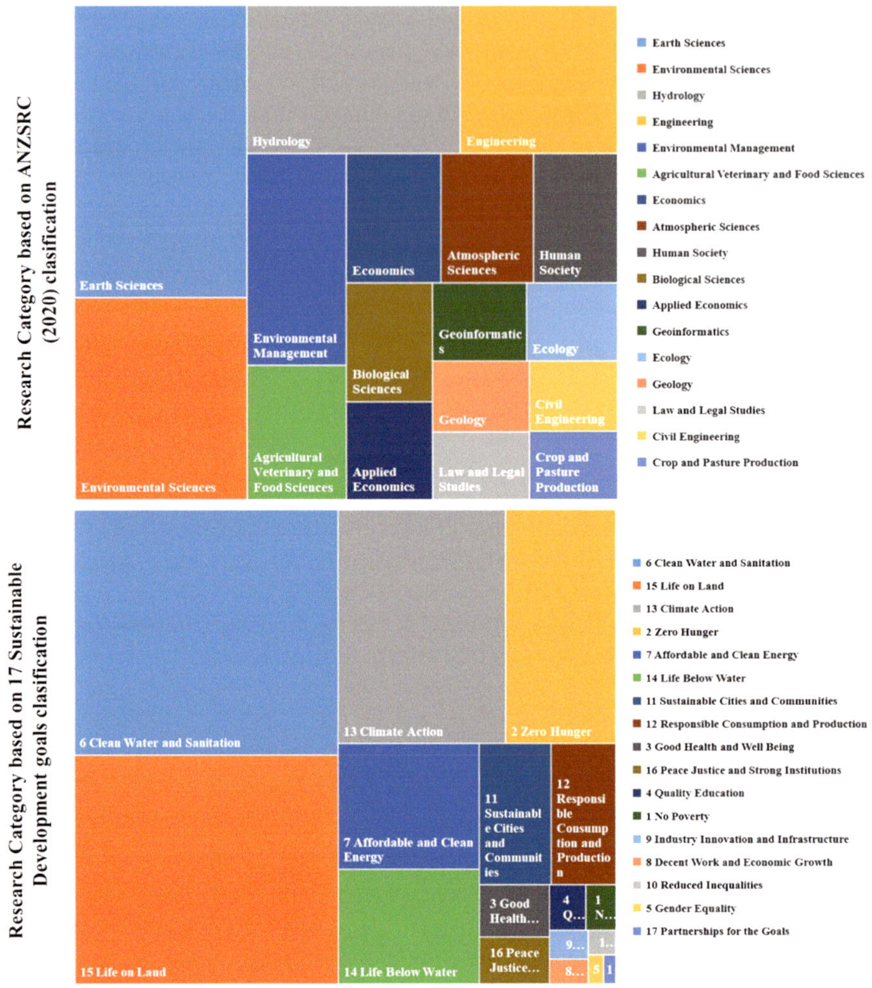

Fig. 3 Tree maps show the classification of research category according to (ANZSRC) and 17SDG (Data retrieved from Dimensions.ai)

water resources, which includes ensuring safe and clean drinking water, as well as safe water for food production and preparation.

- ***SDG 12 Responsible consumption and production*** (2241 documents): This goal also includes responsible consumption of water resources, which is a critical component of water resource management approaches that lead to responsible and sustainable production.
- ***SDG 3 Health and well-being*** (910 documents): The sanitation and accessibility of water is essential for human health and well-being. Accordingly, the achievement of this goal is typically determined by the quality and availability of water resources.

- ***SDG 16 Peace, justice and strong institutions*** (818 documents): It is about providing access to justice for all and supporting peaceful and inclusive societies. In the case of transboundary water disputes, mismanagement of water resources can lead to crises between peoples or countries that share water resources (Rowland 2005). Therefore, sustainable management of water resources is critical to promoting peace and ensuring justice.
- ***SDG 4 Quality education*** (404 documents): Access to water is crucial to ensure quality education in rural areas, especially in developing countries. Therefore, promoting water management would contribute to achieving this goal.
- ***SDG1 Eradicate poverty*** (338 documents): The availability of water resources would prevent the loss of jobs. Many sectors are highly dependent on water quality and availability, so adequate and sustainable management of water resources is essential to achieve SDG1.
- ***SDG9 Industrial innovation and infrastructure*** (278 documents): A strong economy depends on a healthy environment. Thus, green infrastructure includes clean energy, water, and climate resilience, which help to preserve natural environment and support economic development (Global Affairs Canada 2018). Therefore, water management is also critical to support sustainable and inclusive industrialization and promote innovation.
- ***SDG8 Decent work and economic growth*** (239 documents): Many economic activities depend on water, including agriculture, which consumes between 41 and 90% of water resources. Thus, water management is crucial to ensure sustainable growth of economic sectors and thus create jobs.
- ***SDG10 Reduce inequalities*** (167 documents): Sustainable management of water resources can contribute to reducing inequalities, in particular millions of people around the world lack access to good hygiene due to lack of clean water and adequate sanitation.
- ***SDG5 Gender equality*** (107 documents): In many developing countries, access to water is not easy, especially in rural areas where girls and women take on the responsibility of fetching water instead of going to school. This leads to gender inequality. Therefore, safe water management is critical to achieving SDG5.
- ***SDG17 Partnerships for the goals*** (88 documents): Strong and effective partnerships between different stakeholders, including governments, the public and private sectors, and civil society for sustainable water resources management can help promote the implementation and execution of plans and strategies related to water resources management.

The analysis of publications related to WRM based on the two classifications indicated that the sustainable and effective management of water resources:

- At the scientific level must be based on interdisciplinary research projects.
- At the social, economic, and strategic levels, the 17 SDGs must be taken into account when establishing water resources management projects.

For the search query "water resource pollution," similar trends were found based on the 17 SDGs, while based on the ANZSRC classification, the field of environmental sciences ranked first with 10,568 documents, followed by earth sciences (6,489), engineering (6,095), environmental management (5,754), and hydrology (3,312).

3.1.3 Evolution of Publications Rates

The analysis of publication trends reveals a consistent increase in research related to 'water resource management' and 'water resource pollution' over the years. The number of documents related to water resource management increased significantly during the 1980s, exceeding the threshold of 100 documents. It is worth noting that the International Water Conference of 1977, held in Mar de Plata, called for coordination within the water sector and recommended expanding agricultural irrigation. However, it did not consider the increase in water demand and related environmental impacts (UN 1977). This explains the increasing trend of the second search query, 'Water resource pollution.' By 1994, over 100 publications were related to the search query 'water pollution'. Both vertical profiles (Fig. 4) show a similar increasing trend, while the number of publications related to water resource management is more significant than those related to water pollution. Therefore, the current focus is on developing strategies and approaches for the effective management of water resources. Water resource management is a challenging task due to the many factors that affect it. To establish effective and operative strategies, collaboration and partnership between different stakeholders, including researchers from various disciplines, governments, private and public sectors, farmers, and civil society, must be increased and strengthened.

3.2 Bibliometric Analysis Using VosViewer

3.2.1 Network Map

500 documents were downloaded from Dimensions' database and used as input data to construct a network map of the main keywords related to our search query. The publications were published between 2000 and 2024 and sorted by relevance to the search query. Out of 13,692 terms, only 295 met the threshold where the minimum number of occurrences of a term is 10. The most relevant terms based on the calculated relevance score are 170. Figure 5 displays the network visualization map of 170 terms that are most relevant to the search query 'Water resource management'. The figure shows four clusters linked by 9,086 links and 19,562 total link strength. The network visualization map is based on the weight of keyword or term occurrence.

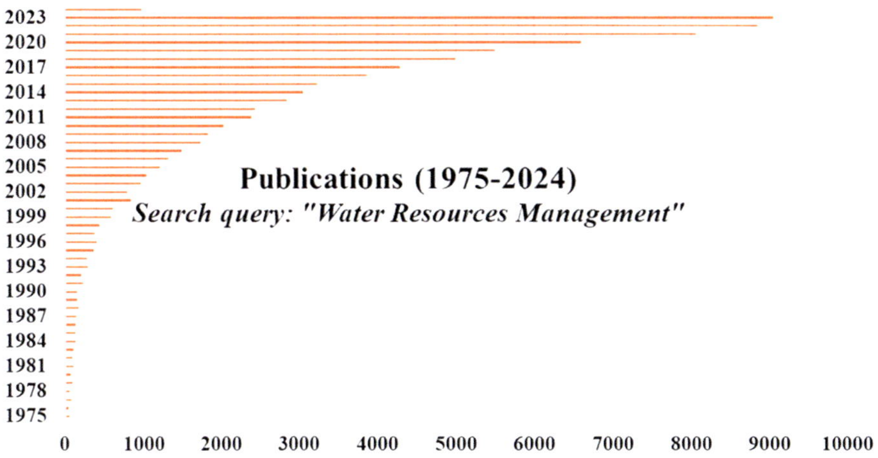

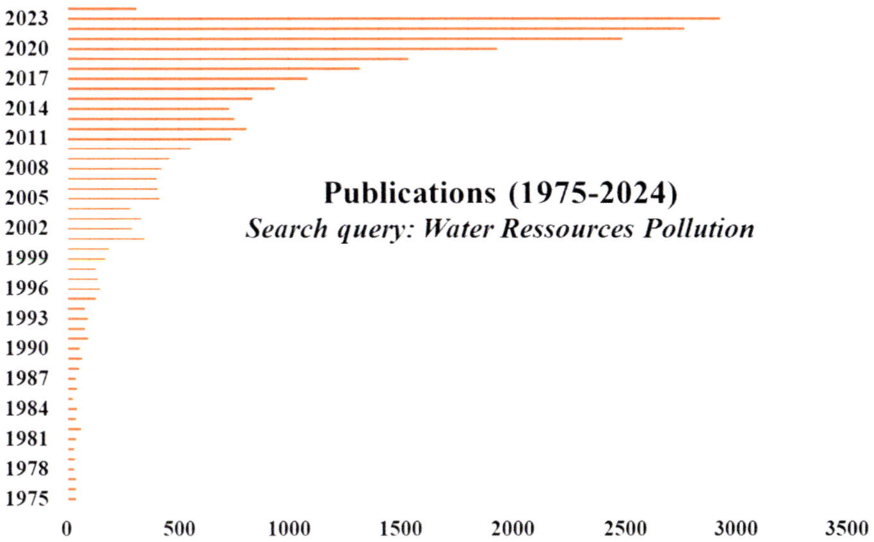

Fig. 4 Number of publications versus years (Retrieved from dimensions)

The cluster in red contains 66 items or keywords. with the most frequent keywords being "issue", "implementation", "technology", "treatment", "energy" and "concept", which occur 62, 52, 50, 48, 48 and 40 times respectively. This cluster pertains primarily to publications on the implementation of approaches and technologies that aid in water governance and management, particularly through the reuse of wastewater to address water stress. The use of terms such as 'gap', 'lack', and 'limitations'

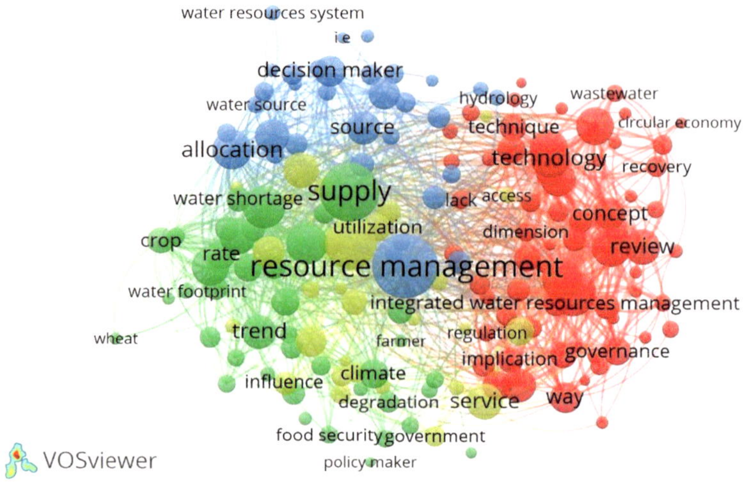

Fig. 5 Map of network visualization based on terms/keywords occurrence weight

indicates the challenges that exist in implementing and applying wastewater treatment systems for society (Morris et al. 2021). Many studies have addressed the challenges related to wastewater management (Jaramillo and Restrepo 2017; Adhikari and Halden 2022; Onu et al. 2023; Silva 2023; Al-Hazmi et al. 2023).

The cluster in green consists of 41 keywords. This cluster covers keywords related to supply (occurrences: 83), irrigation (occurrences: 46), climate (occurrences: 35), crop (occurrences: 30), drought (occurrences: 33), water use (occurrences: 29), food (occurrences: 28), water scarcity (occurrences: 28), and food security (occurrences: 20). It covers topics related to climate variations and their impact on agriculture, crop productivity, and food security.

The blue cluster contains 33 keywords related to hydrology, which is the study of water distribution and movement in the Earth's systems. Hydrology aims to understand these systems and help solve water-related issues. The yellow cluster is related to the impacts of human activities on water quality and the monitoring of water pollution. The use of keywords such as 'economy' and 'economic development' indicates the economic significance of water, as it contributes to the global economy both directly and indirectly. Therefore, water scarcity caused by pollution has a negative impact on economic growth (Distefano and Scott 2017).

3.2.2 Overlay Map

The map visualizes keywords based on their occurrence weight and average publication year (Fig. 6). The early research period (average publication year: 2016–2018) focused on water resource management, including reuse and integrated water resource management. Subsequent research (average publication year: 2018–2020)

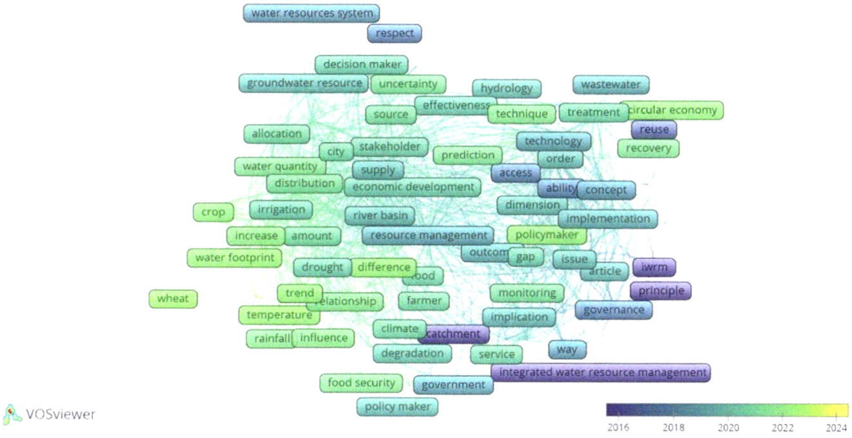

Fig. 6 An overlay visualization map. The map displays average publication year by terms or keyword: purple for keywords adopted to index research studies with the average publication year around 2016, dark green for 2016, clear green for 2020–2022 and yellow–green for 2022–2024

explored the relationship between water resource management and economic development. Research topics related to technologies and techniques for wastewater treatment and reuse are associated with an average publication year between 2020 and 2022. Additionally, the map displays the latest areas of research related to water resource management, including the impact of climate variation on agriculture (Corresponding keywords such as rainfall, temperature, precipitation, influence, and crop). Accordingly, the time map of keywords related to the search query shows an earlier research interest in integrated water resource management and reuse, as well as technologies and approaches for wastewater treatment. The most recent research study direction focuses on the effects of climate change on economic activities, mainly agriculture.

3.2.3 Density or Heat Map Based on Bibliographic Data: Co-authorship

To analyze the distribution of publication trends among countries, we generated a density map based on bibliographical data (co-authorship). We excluded countries with fewer than 10 publications. Out of 114 countries, only 50 met the threshold. The countries with the highest number of publications, China and the United States, are shown as hotspots in warm colors, which is in good accordance with Li et al.'s (2020) findings. Other countries with a significant number of publications include the United Kingdom, Germany, Australia, and Iran. Countries with fewer publications, such as Chile, Argentina, Ireland, Taiwan, and Czechia, are shown in cooler colors. The heat map displays a very irregular distribution of publications across the world, where developing countries are not well shown on the map. The issue of water shortage is

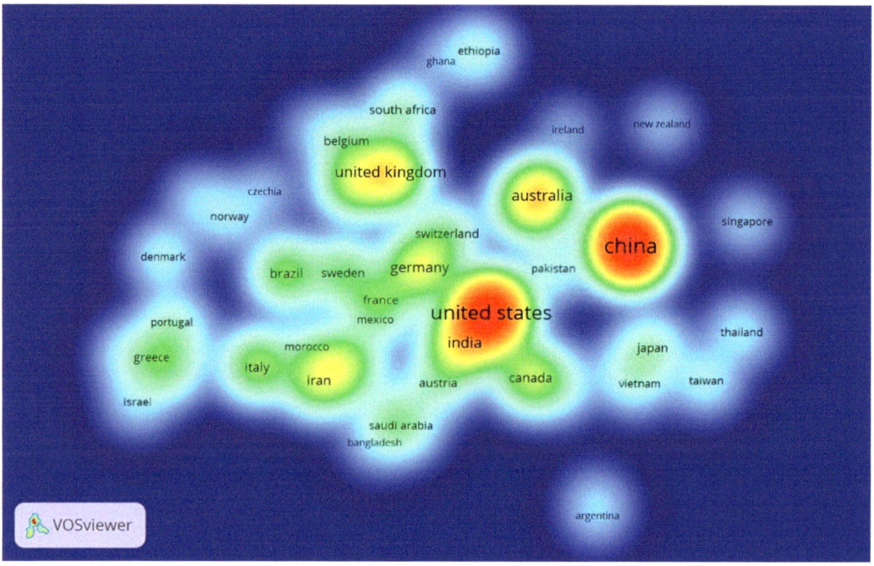

Fig. 7 Heat map based on bibliographic data: co-authorship (density visualization)

a significant global concern. Many studies have reported that developing countries suffer from water scarcity (Vairavamoorthy et al.2008; Starkey 2012; Emile et al. 2022). However, there is a lack of research on water resource management in these countries. Therefore, future studies should focus on countries experiencing water scarcity.

The overlay visualization of the number of publications by country based on co-authorship data shows that the leading countries in the field of WRM in 2016 were the United States, the United Kingdom, and Australia (Fig. 8). In 2018, there was a slight increase in publications in countries such as France, South Africa, and Mexico. Currently, publications related to WRM are being published by researchers from China, which is the leading country in the field (Fig. 7), mainly between 2020 and 2022. However, between 2022 and 2024, other countries such as Ethiopia, Morocco, Pakistan, and Vietnam, are shown on the time map, designating an increased interest in the field of water resource management research.

4 Discussions and Conclusions

Nowadays, the availability and quality of water resources are major concerns facing countries worldwide. Developing strategies and approaches for water resource conservation and protection is crucial for future generations. However, the exploitation and use of water resources vary from country to country and depend on the human

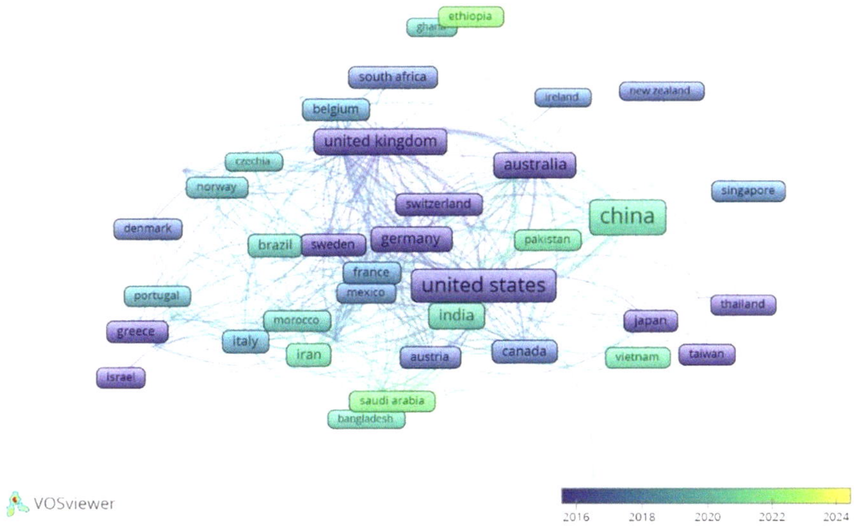

Fig. 8 An overlay visualization map. The map displays average publication year by country: purple for keywords adopted to index research studies with the average publication year around 2016, dark green for 2016, clear green for 2020–2022 and yellow–green for 2022–2024

activities that promote the country's economy. This indeed made the management of water resources challenging.

The effects of climate change and the irrational and irregular exploitation of water resources have made it difficult to adapt and apply management strategies for water resources.

This study analyzes research on water resource management and water resource pollution. Based on the ANZSRC and 17 SDGs classifications, current research on water resource management is primarily focused on environmental sciences, earth sciences, engineering, and environmental management and hydrology. Therefore, there is an urgent need to develop studies on approaches and strategies to address the issue of water resource shortages. In order to achieve sustainable and effective management of water resources, it is necessary to integrate interdisciplinary research projects at scientific, societal, economic, and strategic levels. It is important to consider all 17 SDGs when establishing water resources management projects.

The bibliometric analysis indicates that current research on water resources management is increasingly focused on climate variability and its impact on agricultural activities and productivity. Water scarcity and quality degradation have ecological, societal, and economic impacts, particularly on the agricultural sector. Therefore, the focus is on developing strategies and approaches for water management that maintain both water and food security.

Research studies on water resources management are concentrated in China and the United States. Out of 114 countries, only 50 have published more than 10 papers in this area, indicating an uneven distribution of research. It is worth noting that the number of studies reporting threats of water shortage in developing countries remains insignificant compared to those in developed countries. Water resource management must be a priority in these countries, as agricultural activities are the main sector of the developing country's economy and consume between 70 and 90% of water resources.

References

Adhikari S, Halden RU (2022) Opportunities and limits of wastewater-based epidemiology for tracking global health and attainment of UN sustainable development goals. Environ Int 163:107217

Al-Hazmi H E, Mohammadi A, Hejna A, Majtacz J, Esmaeili A, Habibzadeh S, Saeb MR, Badawi M, Lima EC, Mąkinia J (2023) Wastewater treatment for reuse in agriculture: prospects and challenges. Environ Res 116711

Bruneau J (2009) Human activities and water use in the south saskatchewan river basin joel bruneau, darrell r. corkal, elise pietroniro, brenda toth, and garth van der kamp. A dry oasis: institutional adaptation to climate on the Canadian Plains, vol 24, p 129

Cech TV (2018) Principles of water resources: history, development, management, and policy. Wiley

Chen H, Liu H, Chen X, Qiao Y (2020) Analysis on impacts of hydro- climatic changes and human activities on available water changes in Central Asia. Sci Total Environ 737:139779

Dey P, Mishra A (2017) Separating the impacts of climate change and human activities on streamflow: a review of methodologies and critical assumptions. J Hydrol 548:278–290

Distefano T, Scott K (2017) Are we in deep water? Water scarcity and its limits to economic growth. Ecol Econ 142:130–147. https://doi.org/10.1016/j.ecolecon.2017.06.019

Doungmanee P (2016) The nexus of agricultural water use and economic development level. Kasetsart J Soc Sci 37(1):38–45

Emile R, Clammer JR, Jayaswal P et al (2022) Addressing water scarcity in developing country contexts: a socio-cultural approach. Humanit Soc Sci Commun 9:144. https://doi.org/10.1057/s41599-022-01140-5

Gleick PH, Heberger M (2014) Water and conflict. The world's water 159–171

Global Affairs Canada (2018) Canada's implementation of the 2030 agenda for sustainable development: voluntary national review

Huo Z, Feng S, Kang S, Li W, Chen S (2008) Effect of climate changes and water-related human activities on annual stream flows of the Shiyang river basin in arid north-west China. Hydrol Process: Int J 22(16):3155–3167

Jaramillo MF, Restrepo I (2017) Wastewater reuse in agriculture: a review about its limitations and benefits. Sustainability 9(10):1734

Karan E, Asgari S (2021) Resilience of food, energy, and water systems to a sudden labor shortage. Environ Syst Decis 41(1):63–81

Li Q, Guo X, Zhang L (2020) Bibliometric analysis of water resource management. In: Hu C, Cai M (eds) Geo-informatics and Oceanography. J Coast Res. Special Issue No. 105. Coconut Creek (Florida), pp 210–214. ISSN 0749-0208

Mancosu N, Snyder RL, Kyriakakis G, Spano D (2015) Water scarcity and future challenges for food production. Water 7(3):975–992

Mejjad N, Fekri A, El Hammoumi O, El Aouidi S, El Kharraz J, Kaya S, Moumen A (2021) Application of DPSIR framework to analyze the groundwater pollution threats of municipal solid waste: case study Médiouna Landfill, Morocco. In: E3S web of conferences, vol 314. EDP Sciences, p 06004

Mohamed NN (2019) Egyptian food insecurity under water shortage and its socioeconomic impacts. Conv Water Resour Agric Egypt 245–273

Mohammadi Ghaleni M, Ebrahimi K (2015) Effects of human activities and climate variability on water resources in the Saveh plain, Iran. Environ Monit Assess 187:1–17

Morris JC, Georgiou I, Guenther E et al (2021) Barriers in implementation of wastewater reuse: identifying the way forward in closing the loop. Circ Econ Sustain 1:413–433. https://doi.org/10.1007/s43615-021-00018-z

Onu MA, Ayeleru OO, Oboirien B, Olubambi PA (2023) Challenges of wastewater generation and management in sub-Saharan Africa: a review. Environ Chall 100686

Report of the UN Water Conference (1977). https://www.geo.fu-berlin.de/en/v/iwrm/Introduction/History-of-IWRM/Bilder-History/UN-1977.png?html=1&locale=en&ref=64493359. Accessed 09 Feb 2024

Ripoll J, Urban L, Staudt M, Lopez-Lauri F, Bidel LP, Bertin N (2014) Water shortage and quality of fleshy fruits—making the most of the unavoidable. J Exp Bot 65(15):4097–4117

Ritchie H, Roser M (2023) Water use and stress. Our World in Data

Rosegrant MW, Ringler C, Zhu T (2009) Water for agriculture: maintaining food security under growing scarcity. Annu Rev Environ Resour 34:205–222

Rowland M (2005) A framework for resolving the transboundary water allocation conflict conundrum. Groundwater 43(5):700–705

Silva JA (2023) Wastewater treatment and reuse for sustainable water resources management: a systematic literature review. Sustainability 15(14):10940

Starkey M (2012) The water crisis in third world countries. https://digitalcommons.liberty.edu/cgi/viewcontent.cgi?article=1301&context=honors. Accessed 11 Feb 2024

Sukanya S, Joseph S (2023) Climate change impacts on water resources: an overview. In: Visualization techniques for climate change with machine learning and artificial intelligence, pp 55–76

Taka M, Ahopelto L, Fallon A, Heino M, Kallio M, Kinnunen P, Niva V, Varis O (2021) The potential of water security in leveraging agenda 2030. One Earth 4(2):258–268

UNICEF (2023) Water scarcity. https://www.unicef.org/wash/water-scarcity. Accessed 05 Feb 2024

Vairavamoorthy K, Gorantiwar SD, Pathirana A (2008) Managing urban water supplies in developing countries–climate change and water scarcity scenarios. Phys Chem Earth Parts a/b/c 33(5):330–339

Wang W, Shi Y, Zhang C, Hong S, Shi L, Chang J, Jin Y, Zhuo S, Siang OC, Li R, Wang P (2019) Simultaneous production of fresh water and electricity via multistage solar photovoltaic membrane distillation. Nat Commun 10(1):3012

Xia J, Qiu B, Pan X, Weng J, Fu G, Ouya R (2012) Assessment of water resources vulnerability under climate change and human activities. Adv Earth Sci 27(4):443

Overview of the Evolution of Marine Intrusion Research from 2000 to 2022

Yahya El Hammoudani, Fouad Dimane, Khadija Haboubi, Lahcen Benaabidate, Abdelhak Bourjila, Chaimae Benaissa, Iliass Achoukhi, Abdelaziz Touzani, Sara Bouhout, Hatim Faiz, Aouatif El Abdouni, and Chaimae Haboubi

Abstract The infiltration of saltwater into groundwater sources along coastal areas can cause a substantial rise in salinity, impacting the main water supply in these areas. The purpose of this paper is to deliver a comprehensive examination of contemporary research patterns in groundwater intrusion. The study concentrates on the period spanning 2000–2022, and it involved analyzing 1908 papers from the Scopus database utilizing the '*bibliometrix*' package in the R programming language. The study's findings, which were visualized using '*biblioshiny*' and '*VOSviewer*', demonstrate that research on marine intrusion is on the rise. Interdisciplinary teams of researchers from various countries are working together to attain integrated outcomes. As evidenced by the investigation, the utilization of spatial data science and remote sensing has experienced a significant surge in this field. This paper aims to provide valuable insights to inspire researchers by presenting the current state of research on this topic, including popular research methods, the involvement of authors, and directions for future research.

Keywords Authorship · Bibliometric analysis · Marine intrusion · Scopus

---

Y. El Hammoudani (✉) · F. Dimane · K. Haboubi · A. Bourjila · I. Achoukhi · A. Touzani · S. Bouhout · H. Faiz · A. El Abdouni · C. Haboubi
Department of Energy and Environmental Civil Engineering, Engineering Sciences and Applications Laboratory, National School of Applied Sciences of Al-Hoceima, Abdelmalek Essaâdi University, Tetouan, Morocco
e-mail: elhammoudani5@gmail.com

L. Benaabidate
Laboratory of Functional Ecology and Environment Engineering, University of Sidi Mohammed Ben Abdellah, 30500 Fez, Morocco

C. Benaissa
Department of Earth Sciences, Geosciences Research Team on Natural Risks, Faculty of Science and Technology of Tangier, Abdelmalek Essaâdi University, Tetouan, Morocco

N. Rebai et al. (eds.), *GIS, Applied Computing and Data Science for Water Management*, Lecture Notes in Geoinformation and Cartography,
https://doi.org/10.1007/978-3-031-63038-5_3

1 Introduction

Typically, human activities, especially agriculture, are attracted to locations near waterways and even more so near the ocean. These coastal areas are highly desirable, especially in dry and semi-arid countries. The coastal plains, filled with alluvial soil, have underground water reservoirs with high water levels close to the surface, creating fertile soil (Agoubi 2021; Elabdouni et al. 2022).

Groundwater is a highly solicited water reserve. It is pure water, normally protected by the ground that covers it. When pollution penetrates it, especially by massive infiltration, it is more or less delayed, and not direct and rapid as in the case of surface water. On the other hand, the contamination takes much longer to disappear because of the slow circulation of groundwater. In the case of karst terrain, the characteristics of the pollution may be the same to as the surface water (Bourjila et al. 2021; Taher et al. 2022).

Over the past few years, the utilization of groundwater resources has intensified to cater to the escalating demand for water, which is propelled by heightened socio-economic activities and the rapid expansion of the population (Hajji et al. 2022; Hounsinou 2020). The persistent reduction in water table levels and the concurrent escalation in salinity levels are attributable to multiple factors, including but not limited to the incursion of seawater into coastal aquifers, utilization of agrochemicals such as fertilizers and pesticides, among others (Elabdouni et al. 2020; Abdouni et al. 2021).

The occurrence of seawater intrusion is a significant contributor to elevated salinity levels observed in groundwater utilized for potable and agricultural purposes in coastal regions. In specific scenarios, saline water penetrates inland, thereby polluting the underlying coastal aquifers (Bourjila et al. 2020; Benaissa et al. 2020). In the absence of disturbance (e.g. overexploitation), the penetration of sea level inland is naturally governed by a difference in density (25 kg/m^3) between two waters: one salty and marine, denser and therefore more subject to gravitational forces; the other more or less “fresh” coming from the infiltration of precipitation and thus constituting a coastal water table flowing to the sea (Hammoudani and Dimane 2020a). The demarcation between the freshwater and saltwater bodies occurs abruptly within a narrow region spanning no more than a few meters. This boundary region, commonly referred to as a sharp interface, separates the two homogeneous fluids and often manifests as a saltwater wedge whose inclination slopes towards the continent (Benaissa et al. 2022, 2023).

An examination of relevant and informative published articles is necessary to identify research perspectives pertaining to marine intrusion (Shammi et al. 2019; Zghibi et al. 2011, 2016). The field has undergone tremendous transformation in recent years, with the amalgamation of several related topics and advancements in numerical computing and the widespread use of the internet (Bourjila et al. 2020, 2021, 2023; Ferchichi et al. 2018). The proliferation of papers on the subject has boosted the efficiency of marine intrusion research through comprehensive review articles (Purwoarminta et al. 2018). Typically, the dissemination of studies is hindered by a conventional approach. To date, no alternative method such as a bibliometric analysis examination of marine intrusion evaluation has been explored (Benaissa et al. 2022; Hammoudani and Dimane 2020b, 2021; Hammoudani et al. 2019; Dimane and Hammoudani 2021).

Bibliometric analysis is a technique employed to provide an overview and delineate the developing patterns within a particular domain of research or subject matter. This approach is particularly useful for broad and significant topics that would be challenging to review manually. Bibliometrics has gained widespread use in commercial research and is used to analyze research articles quantitatively and qualitatively to evaluate the development of the topic in the scientific field (Donthu et al. 2021; Moral-Muñoz et al. 2020). Researchers across various fields use bibliometric analysis to track the evolution of articles, collaborations, journal performance, intellectual structures, and social structures. The analysis involves examining the primary metadata of articles and interpreting it with a statistical approach. The results of the analysis are presented in an attractive and easy-to-understand visual representation, providing a more comprehensive view of the subject compared to traditional methods (Aria and Cuccurullo 2017; Aria et al. 2017).

The current pool of review articles is insufficient in characterizing the trends and advancements within the domain of marine intrusion research. As a result, this study endeavors to employ bibliometric analysis to identify the progression of marine intrusion research over the past twenty-two years. To the best of the author's knowledge, this approach has yet to be utilized for recognizing the evolution of marine intrusion via bibliometric analysis. The research primarily focuses on the period between 2000 and 2022 to extract the core of the past and current conditions. The objectives of this study encompass quantifying the cumulative information obtained from published works, identifying pertinent sources that reflect the intricacies of this topic, recognizing key authors who have significantly contributed to the field, analyzing the emerging keywords and subtopics that are most salient, developing a network analysis to comprehend the connections and variations between authors and co-author papers, and delineating the scientific output by country and overall collaboration between countries. The study draws upon research articles retrieved from the renowned Scopus database.

This work can provide several impacts, such as:

- Identifying research gaps in the field of marine intrusion where research is lacking or where further investigation is needed.
- Assessing the impact of research on marine intrusion. By analyzing citation counts and other metrics, researchers can identify the most influential papers, authors, and institutions in the field, and can evaluate the impact of individual research projects.
- Tracking research trends in the field of marine intrusion. By analyzing publication patterns and keyword usage, researchers can identify emerging research areas and track changes in research focus over time.
- It can provide valuable evidence to inform policy decisions related to marine intrusion. By identifying the most impactful research and the most active researchers

in the field, policymakers can make informed decisions about resource allocation and research priorities.

- Promoting collaboration among researchers working on marine intrusion. By identifying researchers with similar interests and expertise, it can help researchers find potential collaborators and establish research partnerships.

Overall, this work can provide valuable insights into the research landscape in the field, and can help inform research priorities, policy decisions, and collaboration efforts.

2 Methodological Approach

Collecting data is a crucial step in conducting a bibliometric analysis. To accomplish this task, a number of databases are available online, all of which are globally recognized and authenticated. The data collection process was performed with largest database of peer-reviewed literature abstracts and citations, Scopus. The database in question offers a comprehensive overview of the worldwide research output, encompassing a diverse range of sources such as scientific journals, books, and conference proceedings. To obtain insight into recent patterns in marine intrusion research, it was crucial to undertake a literature search utilizing specific criteria. These criteria included searching for articles with the terms "marine" and "intrusion" in their titles, abstracts, or keywords, while limiting the publication years to 2000–2022. Additionally, the search was narrowed down to articles classified under subject areas such as Earth Sciences, Environmental Science, Agriculture, Chemistry, and Chemical Engineering. Finally, the language of the articles was restricted to English.

The process of conducting a bibliometric analysis involves several steps, starting with data collection. The data collected was filtered to include only published research articles and documents written in English. The initial search results yielded over 2,890 documents, but after filtration, only 1,908 were selected for analysis. This search was conducted on February 2nd, 2023 (Fig. 1).

The next steps in a bibliometric analysis include filtering the data, analyzing it, visualizing it, and interpreting the results. The analysis considers various parameters such as the number of sources, documents, and average citations per document, keywords used, and authors involved in the research. It also examines the relationships between the different components of the search. Furthermore, in-depth analysis of the collected data may require the use of techniques such as co-citation analysis and bibliographic linkage keyword analysis to elucidate the structural relationships and intellectual interactions between the various articles (Aria and Cuccurullo 2017; Aria et al. 2017).

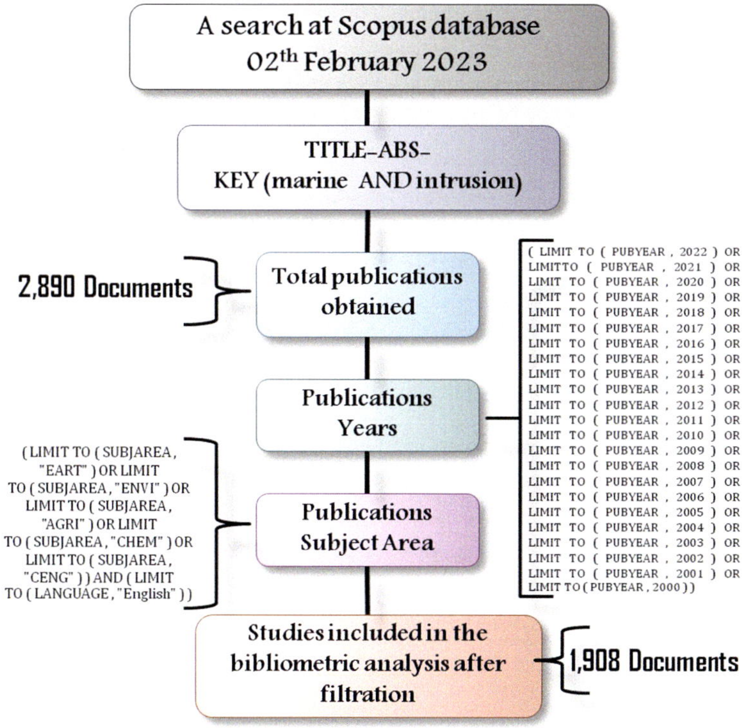

Fig. 1 Summary of the bibliometric analysis

3 Results and Discussions

1,809 research papers focused on marine intrusion as the central theme, with over 7,000 authors contributing to the topic in the last 22 years. While only 132 of these papers were written solely by one author, the majority of them involved collaboration, with an average of over 5 authors per paper (Table 1).

The number of annual scientific publications on marine intrusion has steadily risen over the years, as seen in Fig. 2. In 2020, the year with the highest number of publications, there were 168 papers produced, followed by 157 in 2021. Prior to 2016, the number of annual publications was below 100. The annual growth rate of scientific production is 6.32%. Over time, the average number of citations per year has also increased. Analysis of the topic revealed more than 2,611 keywords and a Scopus machine learning process generated approximately 11,590 words that are closely related. The majority of the papers are the result of collaboration among authors.

Table 1 General informations collected

Description	Results
Main information about data	
Time span	2000–2022
Sources (journals, books, etc.)	654
Documents	1,908
Annual growth rate %	6.32
Document average age	8.95
Average citations per doc	26.35
Document contents	
Keywords plus (ID)	11,590
Author's keywords (DE)	5,636
Authors	
Authors	7,658
Authors of single-authored docs	122
Authors collaboration	
Single-authored docs	132
Co-authors per doc	4.96
International co-authorships %	35.85
Document types	
Article	1,668
Book	3
Book chapter	64
Conference paper	109
Conference review	2
Editorial	1
Erratum	2
Letter	1
Note	2
Retracted	1
Review	55

3.1 Most Relevant Sources

The annual publication trend is a valuable metric for assessing the current state of research in a particular field, as it indicates the level of interest and engagement of researchers in exploring various topics from a broad perspective. In the realm of marine intrusion research, numerous journals (as listed in Table 2) have promoted, reviewed, and published related articles. Among them, ORE GEOLOGY

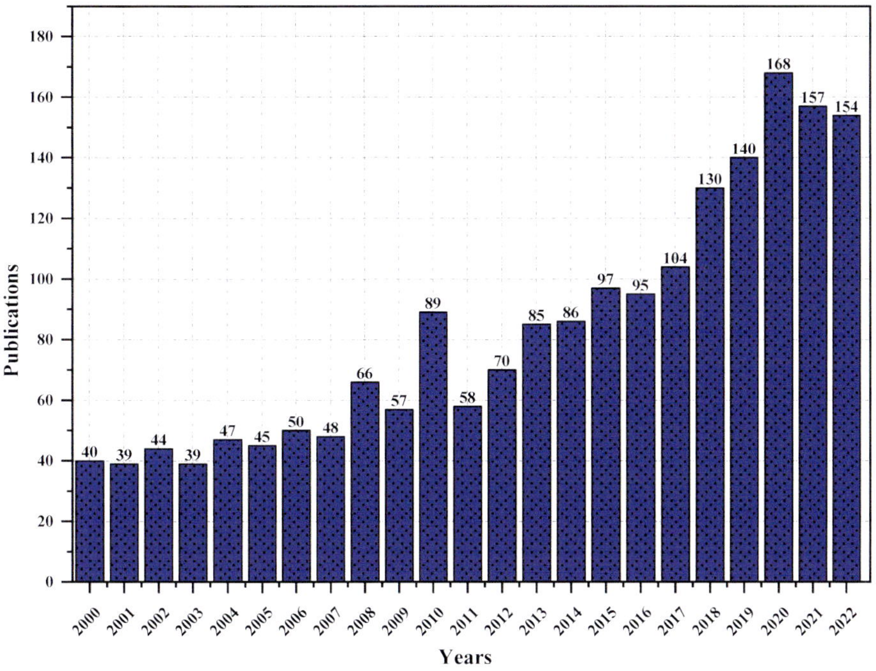

Fig. 2 Productivity on marine intrusion over time

REVIEWS has been the most prolific publisher in the last 22 years, with 29 articles on urban groundwater quality. SCIENCE OF THE TOTAL ENVIRONMENT has also been a significant contributor, followed by ESTUARINE, COASTAL AND SHELF SCIENCE and MARINE AND PETROLEUM GEOLOGY, with 28 and 27 articles respectively. By examining the reference list, it is apparent that SCIENCE OF THE TOTAL ENVIRONMENT is the most frequently cited, with 471,465 citations, followed by JOURNAL OF HYDROLOGY with 51,262 citations. It is worth noting that MARINE AND PETROLEUM GEOLOGY and MARINE POLLUTION BULLETIN have also made notable contributions in terms of publication frequency and citation count over the years. Figure 3 shows journals dynamicity over years.

3.2 *Highest Productive Authors*

In this study, we conducted a bibliometric analysis to identify authors who have made important contributions to the research field by publishing articles related to a concerning topic. Our analysis revealed the top ten most productive authors who have produced a substantial amount of information on the topic, as reflected by their

Table 2 Most relevant sources

Journal	Articles	Total publications	Total citations	Cite score	Most cited article	Times cited	H-index	Publisher
Ore geology reviews	29	2,092	12,589	6.0	Orogenic gold deposits: a proposed classification in the context of their crustal distribution and relationship to other gold deposit types (Groves et al. 1998)	1,539	106	Elsevier
Science of the total environment	28	28,505	471,465	16.5	A review on the occurrence of micropollutants in the aquatic environment and their fate and removal during wastewater treatment (Luo et al. 2014)	2,453	275	Elsevier
Estuarine, coastal and shelf science	27	8,838	1,643	5.4	Mangrove forests: Resilience, protection from tsunamis, and responses to global climate change (Alongi 2008)	1,036	142	Elsevier
Marine and petroleum geology	27	2,115	19,388	9.2	The importance of shale composition and pore structure upon gas storage potential of shale gas reservoirs (Ross and Bustin 2009)	1,780	129	Elsevier

(continued)

Table 2 (continued)

Journal	Articles	Total publications	Total citations	Cite score	Most cited article	Times cited	H-index	Publisher
Marine pollution bulletin	26	4,072	40,169	9.9	Microplastics in the marine environment (Andrady 2011)	3,717	193	Elsevier
Palaeogeography, palaeoclimatology, palaeoecology	24	1,767	10,582	6.0	Present-day South American climate (Garreaud et al. 2009)	1,031	162	Elsevier
Environmental earth sciences	22	2,559	12,882	5.0	Delineation of landslide hazard areas on Penang Island, Malaysia, by using frequency ratio, logistic regression, and artificial neural network models (Pradhan and Lee 2010)	473	130	Springer Nature
Journal of hydrology	22	4,993	51,262	10.3	River flow forecasting through conceptual models part I—A discussion of principles (Nash and Sutcliffe 1970)	16,134	241	Elsevier
Atmospheric environment	21	–	–	–	Parameterization of surface resistances to gaseous dry deposition in regional-scale numerical models (Wesely 2007)	1,741	257	Elsevier
Journal of African earth sciences	21	1,087	4,687	3.9	The Red Sea and Gulf of Aden Basins (Bosworth et al. 2005)	623	83	Elsevier

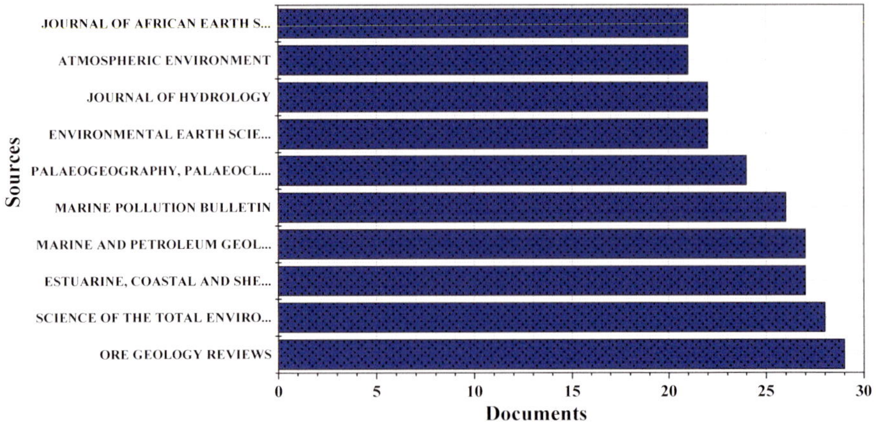

Fig. 3 Production dynamicity over years

H-Index (Figs. 4 and 5). It is worth noting that this topic has attracted the attention of several authors who have reported their observations in their publications.

Among the top ten authors, Professor WANG Y stands out as he has published 26 contributions in the last 22 years and has had a significant impact on the study of marine intrusion with an h-index value of nearly 11. Another noteworthy author is LI X, who has produced 21 papers and has the highest h-index value of 12 among the other authors. LI X's works on groundwater studies and hydrological modeling, as well as his expertise in remote sensing, has had an impact in this research area. ZHANG J from China has presented 18 papers on marine intrusion, while WANG Y, LI X, and CHEN J have collaborated on research related to this topic, which is

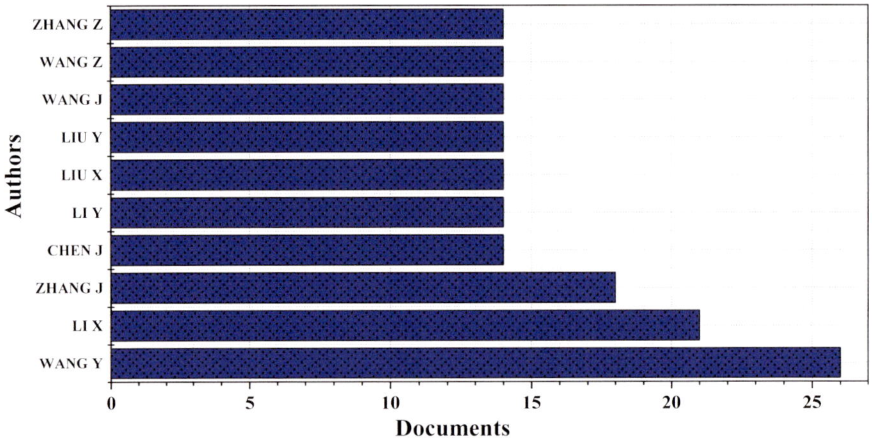

Fig. 4 Most relevant authors

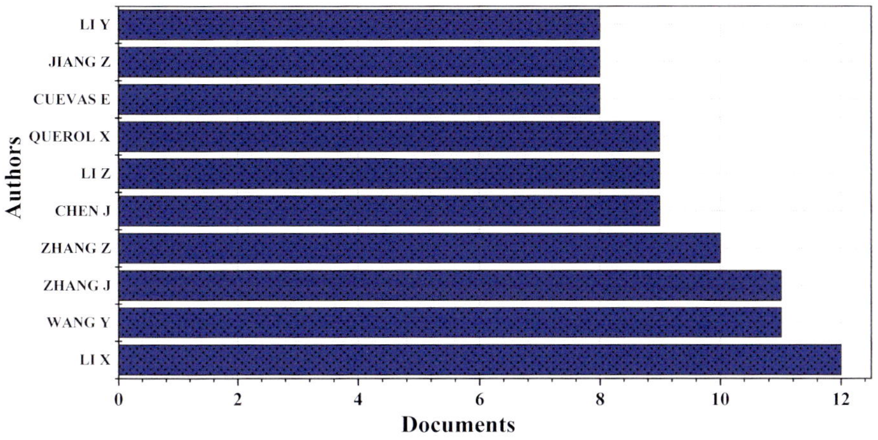

Fig. 5 Authors impact by H-Index

reflected in their publications. CHEN J has published 14 papers, which is the same number as the other authors in the top 10 list.

Lastly, the work of LIU X and LIU Y is noteworthy as they have received a considerable number of citations for their research. They mainly utilized geographic information systems to identify the spatial variation of marine intrusion. Overall, our study has identified the most productive authors and their contributions to the research field on this concerning topic, providing insight into the work that has been done and the direction that the research is heading.

This paper takes a deep dive into the scientific output and citation rates of different countries, and let me tell you, it's pretty eye-opening. Figures 6 and 7 lay it all out, and it turns out that China is killing it with a whopping 1,732 papers and 6,500 citations. The USA isn't far behind with 1,328 papers, but they take the crown for the most citations across all the countries with a jaw-dropping 8,571. And in third place is Australia, putting out 349 papers and racking up a total of 3,896 citations. For the nitty–gritty on how the other countries stack up, scope out Figs. 6 and 7.

3.3 Most Relevant Documents

We conducted a factor analysis to rank the top 10 articles contributing to marine intrusion between 2000 and 2022, based on their total number of citations. The ranking of the most cited articles can be found in Table 3. One of the highly cited articles, Mastalerz et al. (2013), examined the evolution of porosity in shales across a range of maturity levels using five New Albany shale samples. Chough et al. (2000) provided a review of the tectonics and sedimentation of major sedimentary basins and orogenic belts in the Korean Peninsula. Querol et al. (2001) monitored levels

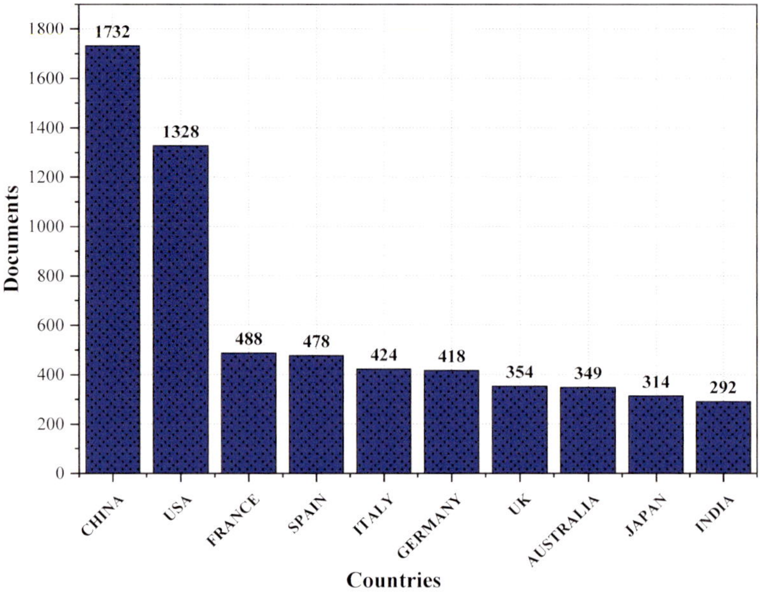

Fig. 6 Highest productive countries

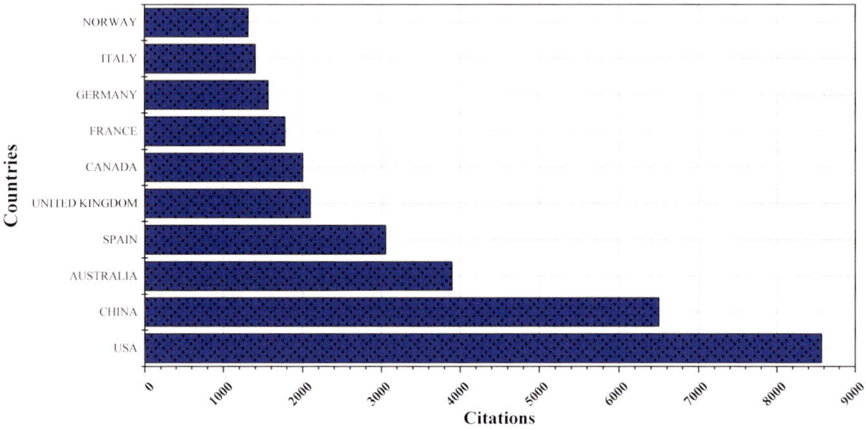

Fig. 7 Most cited countries

of suspended particles on an urban sidewalk in the Barcelona metropolitan area, while Chew et al. (2004) investigated the relationship between microstructure and engineering properties of cement-treated marine clay. Johnson et al. (2011) reviewed the geological events that occurred during the final stages of the development of the northern part of the East African Orogen.

Table 3 Most cited documents

Paper	DOI	Total citations	TC per year	Normalized TC
Mastalerz et al. (2013)	https://doi.org/10.1306/04011312194	662	60.18	16.60
Chough et al. (2000)	https://doi.org/10.1016/S0012-8252(00)00,029-5	622	25.92	8.33
Querol et al. (2001)	https://doi.org/10.1016/S1352-2310(01)00,361-2	541	23.52	8.97
Chew et al. (2004)	https://doi.org/10.1061/(ASCE)1090-0241(2004)130:7(696)	474	23.70	9.08
Johnson et al. (2011)	https://doi.org/10.1016/j.jafrearsci.2011.07.003	435	33.46	10.94
Torsvik and Cocks (2013)	https://doi.org/10.1016/j.gr.2013.06.012	397	36.09	9.95
Preiss (2000)	https://doi.org/10.1016/S0301-9268(99)00,068-6	386	16.08	5.17
Robinson et al. (2007)	https://doi.org/10.1016/j.advwatres.2006.07.006	315	18.53	7.05
Ernst and Buchan (2003)	https://doi.org/10.1146/annurev.earth.31.100901.145500	290	13.81	6.25
Querol et al. (2008)	https://doi.org/10.1016/j.atmosenv.2006.10.071	280	17.50	7.44

The author had a whopping 5,636 keywords in their study, and we went through them to figure out the most commonly used words (check out Figs. 8 and 9). Turns out, 'SEAWATER' (286) and 'SALINE INTRUSION' (269) were the big guns, followed closely by 'GROUNDWATER' (234) and 'SALINITY' (225) in the research. 'CHINA' was also mentioned a ton, around 193 times to be exact. And we also came across some other words that popped up frequently, like 'WATER QUALITY', 'AQUIFERS', and 'ARTICLE'.

3.4 Co-occurrence of Keywords

In Fig. 10, you can see a visual representation of the most commonly used keywords and their co-occurrence network. The dataset used in this study included a total of 1,908 documents, and from these documents, 11,590 keywords were detected (including the author's keywords). To prevent redundancy, the same word was limited to appear a maximum of three times. The keywords were then grouped into six clusters, each represented by a different color (red, yellow, orange, green, blue, and purple).

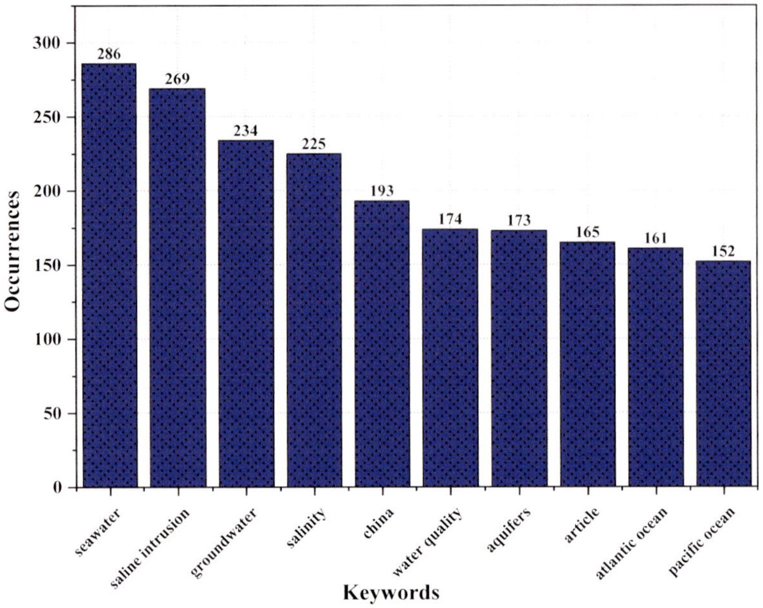

Fig. 8 Most frequent keywords

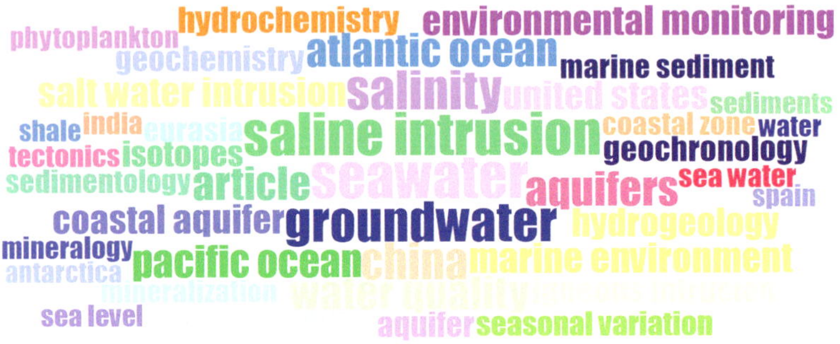

Fig. 9 Word cloud of most relevant keywords

The nodes in the network graph represent individual keywords, and the connections between them indicate that they were found in the same document. Nodes that are closer together indicate that the keywords are more closely related in meaning. Moreover, the more connections a node has, the more frequently the corresponding keyword appears in the articles.

In this study on marine intrusion, the researchers used Scopus data to identify important keywords from the abstracts of related research. Using a machine learning process, these keywords were grouped together into clusters based on their relevance.

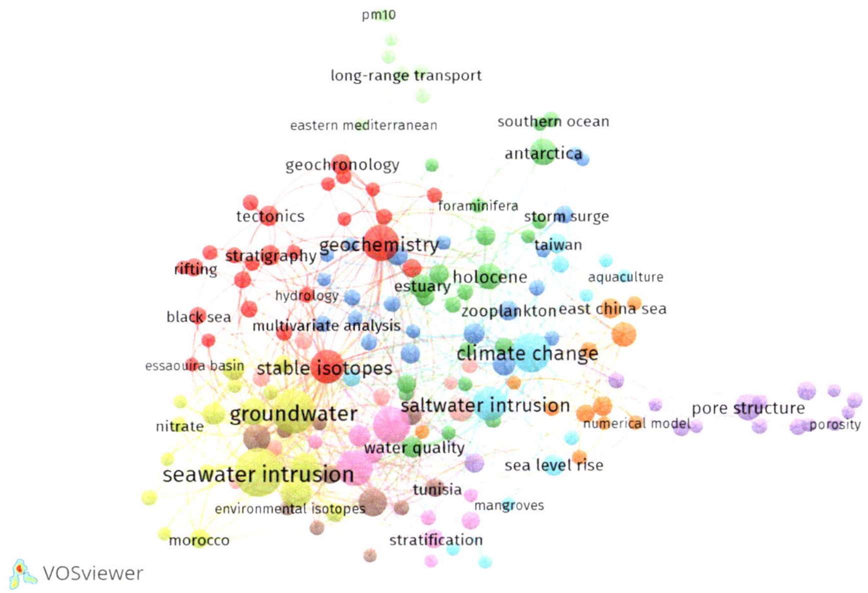

Fig. 10 Co-occurrence of keywords

This helped the researchers gain a better understanding of the topic by organizing the information into themes.

Interestingly, the study found that there were no fundamental themes in this area of research. However, there were a few key themes that emerged as being particularly important. These were "saline intrusion", "seawater", and "groundwater". On the other hand, there were also some emerging themes that the researchers identified. These included "igneous intrusion", "geochronology", and "sedimentology". All of this information is nicely summarized in Fig. 11.

4 Limitations

First of all, this bibliometric analysis can be the starting point for several studies, but it still has certain drawbacks. The Scopus database was the only one used to collect data. As a result, some journals from other databases were not included, such as web of science, Google scholar, PubMed, and Pro-Quest. Using all of the databases may result in different findings.

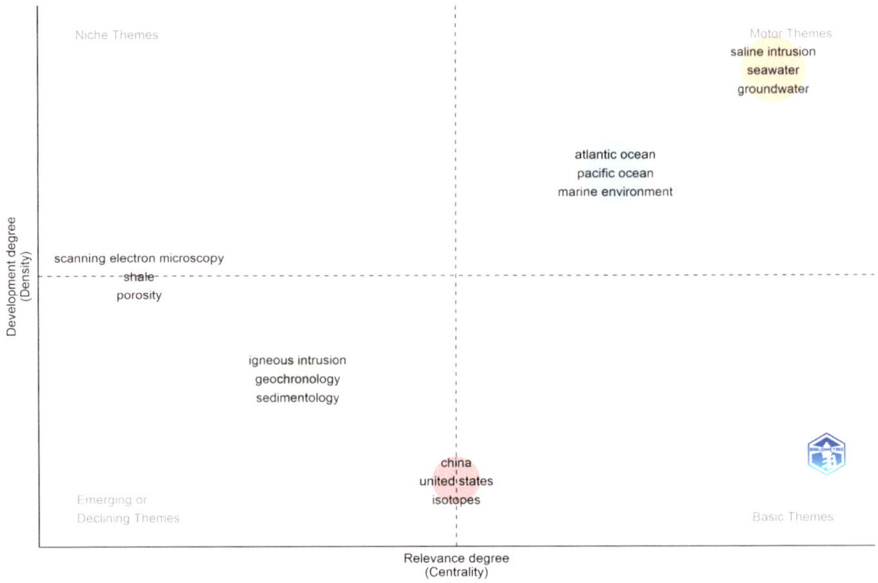

Fig. 11 Thematic map showing significant themes using keyword's plus

5 Conclusions

In this paper, we explored the research patterns related to marine intrusion through the use of bibliometric analysis in the Scopus database. We found that this issue has caused a significant change in groundwater quality, and researchers have identified potential causes and solutions to address it. This area of research presents numerous opportunities for further exploration, given the many aspects of marine intrusion that are still unexplored.

While we acknowledge the limitations of our analysis, which include the possibility of valuable articles being overlooked based on selected parameters and variations in query modifications, we believe that the bibliometric analysis provides a useful quantitative and descriptive approach to understanding the evolution of marine intrusion assessment research over the past 22 years. We also acknowledge that there are drawbacks to relying solely on computing machines, as this can sometimes neglect human experience and perspective in the results.

Overall, we hope that our findings will contribute to a better understanding of the researchers' views on this topic, and that further studies will build on this foundation to better understand and address the issue of marine intrusion.

Competing Interests The author declares that there is no conflict of interests regarding the publication of this manuscript. In addition, the ethical issues, including plagiarism, informed consent, misconduct, data fabrication and/or falsification, double publication and/or submission, and redundancy has been completely observed by the authors.

References

Abdouni AE et al (2021) Physicochemical characterization of wastewater from the Al-Hoceima slaughterhouse in Morocco. Casp J Environ Sci 19(3):423–429

Agoubi B (2021) A review: saltwater intrusion in North Africa's coastal areas—current state and future challenges. Environ Sci Pollut Res 28:17029–17043

Alongi DM (2008) Mangrove forests: resilience, protection from tsunamis, and responses to global climate change. Estuar Coast Shelf Sci 76(1):1–13

Andrady AL (2011) Microplastics in the marine environment. Mar Pollut Bull 62(8):1596–1605

Aria M, Cuccurullo C (2017) Bibliometrix: an R-tool for comprehensive science mapping analysis. J Informetr 11(4):959–975

Aria M, Cuccurullo C, Aria MM (2017) Package 'bibliometrix'

Benaissa C et al (2020) Hydro-chemical and bacteriological Study of Some Sources of Groundwater in the GHIS-NEKOR and the BOKOYA Aquifers (AL HOCEIMA, MOROCCO). In: Proceedings of the 4th edition of international conference on Geo-IT and water resources

Benaissa C et al (2022) Assessment of water quality using water quality index – case study of Bakoya Aquifer, Al Hoceima, Northern Morocco. Ecol Eng & Environ Technol 23(4):31–44

Benaissa C, Bouhmadi B, Rossi A (2023) An assessment of the physicochemical, bacteriological quality of groundwater and the water quality index (WQI) used GIS in Ghis Nekor, Northern Morocco. Sci Afr 20:e01623

Bosworth W, Huchon P, McClay K (2005) The red sea and gulf of aden basins. J Afr Earth Sc 43(1–3):334–378

Bourjila A et al (2020) Use of GIS, remote sensing and AHP techniques to delineate groundwater potential zones in the Nekor basin, Central Rif of Morocco. In: Proceedings of the 4th edition of international conference on Geo-IT and water resources 2020, Geo-IT and Water Resources 2020

Bourjila A et al (2021) Groundwater potential zones mapping by applying GIS, remote sensing and multi-criteria decision analysis in the Ghiss basin, Northern Morocco. Groundw Sustain Dev 15:100693

Bourjila A et al (2023) Mapping the spatiotemporal evolution of seawater intrusion in the Moroccan coastal aquifer of Ghiss-Nekor using GIS-based modeling. Water Cycle 4:104–119

Chew S, Kamruzzaman A, Lee F (2004) Physicochemical and engineering behavior of cement treated clays. J Geotech Geoenviron Eng 130(7):696–706

Chough S et al (2000) Tectonic and sedimentary evolution of the Korean peninsula: a review and new view. Earth Sci Rev 52(1–3):175–235

Dimane F, El Hammoudani Y (2021) Assessment of quality and potential reuse of wastewater treated with conventional activated sludge. Mater Today: Proc 45:7742–7746

Donthu N et al (2021) How to conduct a bibliometric analysis: an overview and guidelines. J Bus Res 133:285–296

El Hammoudani Y, Dimane F (2020a) Assessing behavior and fate of micropollutants during wastewater treatment: Statistical analysis. Environ Eng Res

El Hammoudani Y, Dimane F (2020b) Assessing behavior and fate of micropollutants during wastewater treatment: Statistical analysis. Environ Eng Res 26(5)

El Hammoudani Y, Dimane F (2021) Occurrence and fate of micropollutants during sludge treatment: case of Al-Hoceima WWTP, Morocco. Environ Chall 5(1):1–8

El Hammoudani Y, Dimane F, El Ouarghi H (2019) Fate of Selected Heavy Metals in a Biological Wastewater Treatment System. in Euro-Mediterranean Conference for Environmental Integration. Springer

Elabdouni A et al (2020) Olive mill wastewater (OMW) production in the province of Al-Hoceima (Morocco) and their physico-chemical characterization by mill types. Mater Today: Proc 27:3145–3150

Elabdouni A et al (2022) Removal of organic matter and polyphenols in the olive oil mill wastewater by coagulation-flocculation using aluminum sulfate and lime. Moroccan J Chem 10(1):191–202

Ernst RE, Buchan KL (2003) Recognizing mantle plumes in the geological record. Annu Rev Earth Planet Sci 31(1):469–523

Ferchichi H et al (2018) Assessment of groundwater salinity using GIS and multivariate statistics in a coastal Mediterranean aquifer. Int J Environ Sci Technol 15:2473–2492

Garreaud RD et al (2009) Present-day south American climate. Palaeogeogr Palaeoclimatol Palaeoecol 281(3–4):180–195

Groves DI et al (1998) Orogenic gold deposits: a proposed classification in the context of their crustal distribution and relationship to other gold deposit types. Ore Geol Rev 13(1–5):7–27

Hajji S et al (2022) Assessment of seawater intrusion in coastal aquifers using multivariate statistical analyses and hydrochemical facies evolution-based model. Int J Environ Res Public Health 19(1):155

Hounsinou SP (2020) Assessment of potential seawater intrusion in a coastal aquifer system at Abomey-Calavi, Benin. Heliyon 6(2):e03173

Johnson P et al (2011) Late Cryogenian-Ediacaran history of the Arabian-Nubian Shield: a review of depositional, plutonic, structural, and tectonic events in the closing stages of the northern East African Orogen. J Afr Earth Sc 61(3):167–232

Luo Y et al (2014) A review on the occurrence of micropollutants in the aquatic environment and their fate and removal during wastewater treatment. Sci Total Environ 473:619–641

Mastalerz M et al (2013) Porosity of Devonian and Mississippian New Albany Shale across a maturation gradient: Insights from organic petrology, gas adsorption, and mercury intrusion. AAPG Bull 97(10):1621–1643

Moral-Muñoz JA et al (2020) Software tools for conducting bibliometric analysis in science: an up-to-date review. Prof Inf 29(1)

Nash JE, Sutcliffe JV (1970) River flow forecasting through conceptual models part I—a discussion of principles. J Hydrol 10(3):282–290

Pradhan B, Lee S (2010) Delineation of landslide hazard areas on Penang Island, Malaysia, by using frequency ratio, logistic regression, and artificial neural network models. Environ Earth Sci 60:1037–1054

Preiss W (2000) The Adelaide Geosyncline of South Australia and its significance in Neoproterozoic continental reconstruction. Precambr Res 100(1–3):21–63

Purwoarminta A, Moosdorf N, Delinom RM (2018) Investigation of groundwater-seawater interactions: a review. In: IOP conference series: earth and environmental science.. IOP Publishing

Querol X et al (2008) PM speciation and sources in Mexico during the MILAGRO-2006 Campaign. Atmos Chem Phys 8(1):111–128

Querol X et al (2001) PM10 and PM2. 5 source apportionment in the Barcelona Metropolitan area, Catalonia, Spain. Atmos Environ 35(36):6407–6419

Robinson C, Li L, Prommer H (2007) Tide-induced recirculation across the aquifer-ocean interface. Water Resour Res 43(7)

Ross DJ, Bustin RM (2009) The importance of shale composition and pore structure upon gas storage potential of shale gas reservoirs. Mar Pet Geol 26(6):916–927

Shammi M et al (2019) Impacts of salinity intrusion in community health: a review of experiences on drinking water sodium from coastal areas of Bangladesh. In: Healthcare. MDPI

Taher M, Mourabit T, Bourjila A, Saadi O, Errahmouni A, El Marzkioui F, El Mousaoui M (2022) An estimation of soil erosion rate hot spots by integrated USLE and GIS methods: a case study of the Ghiss Dam and Basin in Northeastern Morocco. Geomat Environ Eng 2(16):95–110

Torsvik TH, Cocks LRM (2013) Gondwana from top to base in space and time. Gondwana Res 24(3–4):999–1030

Wesely M (2007) Parameterization of surface resistances to gaseous dry deposition in regional-scale numerical models. Atmos Environ 41:52–63

Zghibi A, Zouhri L, Tarhouni J (2011) Groundwater modelling and marine intrusion in the semi-arid systems (Cap-Bon, Tunisia). Hydrol Process 25(11):1822–1836

Zghibi A et al (2016) Groundwater vulnerability analysis of Tunisian coastal aquifer: an application of DRASTIC index method in GIS environment. Groundw Sustain Dev 2:169–181

Climate Change and Water Resources in the Maghreb: What Are the Strategies for Adaptation?

Rachid Ech-Choudany and Hicham Hafid

Abstract In the face of a global food crisis and climate change, ensuring access to water resources becomes a critical concern for human activities. The Maghreb region, including Morocco, Tunisia and Algeria, is particularly exposed to ecological vulnerabilities. The environmental degradation in this region has already led to adverse effects on the well-being of populations and hindered economic development. While demographic projections provide some clarity, the uncertainties surrounding climate change projections persist due to inadequate climatological data, both in terms of quantity and quality. Nonetheless, numerous climate projections for Southern Europe indicate a worrisome trend of increasing temperatures and decreasing precipitation. The Maghreb countries, due to their geographical proximity, are expected to face significant repercussions. This challenge is further compounded by the presence of the hyper-arid Sahara, which exacerbates the impact of climate change. Consequently, the Maghreb states recognize the imminent climate evolution as a major concern. This study aims to present an overview of the current climate evolution projections in the Maghreb region and explore their potential consequences on water resources. It encompasses an analysis of the various aspects of climate change, while emphasizing the necessity of implementing adaptation strategies to mitigate the adverse impacts on water resources.

Keywords Maghreb · Morocco · Tunisia · Algeria · Water resource · Climate change · Adaptation strategies · Ecological vulnerabilities

---

R. Ech-Choudany (✉)
Laboratoire d'Analyse Économique et Modélisation, FSJES-Souissi, Rabat, Marocco
e-mail: rachid.echchoudany@um5s.net.ma

H. Hafid
Institut des Études Africaines, Université Mohammed V de Rabat, Rabat, Marocco

N. Rebai et al. (eds.), *GIS, Applied Computing and Data Science for Water Management*, Lecture Notes in Geoinformation and Cartography,
https://doi.org/10.1007/978-3-031-63038-5_4

1 Introduction

In a Maghreb context marked by an exacerbated risk of water shortages, water resource management has become a major concern of public policies in Africa in general and North Africa in particular in recent years. Indeed, the needs of economic growth and demographic expansion have placed the Maghreb region at the heart of this issue. This situation is further exacerbated by rainfall fluctuations, ongoing climate change and inadequate management of available water resources.

According to the World Bank (2018), Tunisia, Algeria, and Morocco will become global hotspots by the end of the 21st century. Increasing temperatures and reduced precipitation will escalate drought frequency in the region (World Bank 2017). Droogers et al. (2012) discerned that climate changes will account for 22% of water shortages by 2050.

Confronted with tensions arising from heightened water demand and competition among diverse uses, policymakers in Maghreb countries have had to explore alternative solutions beyond constructing more reservoirs and interregional transfers. Starting from the 1990s, the three area's countries have introduced new public water management policies, known as the "water mix." This approach involves combining conventional and unconventional resource mobilization (such as seawater desalination and wastewater purification), adopting water-saving irrigation techniques, and curbing losses in distribution networks.

Central to this approach is a demand management policy, supported by economic instruments like progressive pricing, the application of the "polluter-pays" principle, and the restriction of irrigation water quotas. Concurrently, these countries are establishing commercial strategies that promote the import of products with high virtual water content.

The objective of this paper is to assess the content, scope and sustainability of these mid-term strategies in facing of the current effects of climate change on water resources.

Unfavorable hydrogeographical context, Evolution of precipitation, and the impact of climate change in Maghreb (Morocco, Algeria, Tunisia).

The three North African countries of Morocco, Tunisia and Algeria form part of a region that lies between the Mediterranean Sea to the north, the Sahara to the south and the Atlantic Ocean to the west. It is located in a desert region that covers more than two-thirds of its surface area.

Precipitation, already characterized by significant temporal and spatial variations, is now undergoing substantial impacts due to ongoing climate change. Concurrently, population growth, rapid urbanization, and economic development have significantly increased water demands across all sectors. Consequently, these countries recurrently face a major water crisis (Gana 2006). Over the past decades, the Maghreb region has witnessed a decline in water resources (Thivet 2009). As such, the Maghreb countries are consistently affected by water stress. According to the World Resources Institute (WRI 2022), the Maghreb countries will experience extremely high levels of water stress by 2040. Moreover, in terms of precipitation frequency, rising temperatures are resulting in significant alterations in precipitation patterns, leading to reduced rainfall amounts in the region. This crisis situation manifests in the scarcity and degradation of water resources, alongside an increasing competition among diverse water uses (agriculture, tourism, industry, and domestic use). Indeed, water resources are primarily allocated to agriculture, accounting for over 80, 13% for drinking water supply, and the remainder for industry (Agoumi 2003).

The economies of the countries in this area are heavily dependent on agriculture, particularly rain-fed agriculture (dependent on seasonal rainfall). The agricultural sector contributes around 15% of the Gross Domestic Product (GDP), with employment rates of 38% in Morocco, 16% in Tunisia, and 22% in Algeria (FAO 2017). Moreover, agricultural lands cover half of the Maghreb's area. Non-irrigated agriculture in North Africa, which accounts for 50% of total agriculture, is highly reliant on winter precipitation, in particular on total accumulation of precipitation and the frequency of rainy days during this period (Garcia-Ruiz 2011). However, constrained by limited water resources, this sector experiences significant fluctuations in terms of production and yields due to the variability of precipitation patterns, both intra-annually and interannually.

1.1 Temperature Trends

Over the past three decades, a significant increase in annual temperatures has been observed in the Maghreb region (Vizy 2017). During this period, data on annual and seasonal mean temperatures in North Africa reveal a widespread warming due to the inherent natural variability of this geographic area. Furthermore, it is noted that surface temperature averages experience a rise during the spring and summer seasons in the northern part (Barkhordarian et al. 2012).

In recent years, the Mediterranean region has experienced climate warming at a faster rate than the global average, with a particularly pronounced acceleration around the Maghreb during summers. This situation presents differential risks associated with temperature increases between the Scandinavian region of Europe and the Mediterranean region. For instance, during the summer of 2021, temperatures reached levels well above 45 °C in Morocco, nearing the absolute heat record of 50.3 °C recorded on July 13, 1961, in Smara, in the southern provinces of the Kingdom. In Tunisia, in mid-August 2021, the thermometer climbed to 48 degrees in Tunis, surpassing the previous heat record of 46.8 °C set in 1982. During this period, Beja recorded a temperature of 48.9 °C, making it the fourth hottest location worldwide. In Algeria, during the summer of 2021, temperatures reached up to 48 °C in several provinces, including Tizi-Ouzou, Bejaïa, Annaba, and El-Tarf (Jaldi 2022).

According to projections by the Intergovernmental Panel on Climate Change (IPCC) climate warming in the Maghreb could be more severe than previously anticipated. In a scenario where greenhouse gas emissions do not experience a drastic increase or a significant reduction, temperatures are projected to increase by 3.8 °C by 2100 compared to the early 20th century, with summers experiencing an average increase of 5 °C in the Maghreb. This increase would lead to more frequent, intense, and prolonged heatwaves. Furthermore, the effects of warming would be felt at higher altitudes, resulting in a reduced snow cover duration and snow thickness in the Atlas and Rif Mountains in Morocco since the 1960s. In the most extreme scenario of a temperature increase exceeding 4 °C by 2100, all ski resorts in the coastal town of Ifrane would become nonviable without the use of artificial snow (IPCC 2022).

1.2 *Precipitation Largely Absorbed by Intense Evaporation*

Over the past century, the average annual precipitation in the Maghreb has exhibited a decrease. Notably, over the last thirty years (1979–2008), northern regions have seen a clear reduction in winter and early spring precipitation quantities. However, an analysis of observed data reveals a significant increase in autumn precipitation in certain areas of northern Algeria and Morocco (Barkhordarian et al. 2013).

In general, it has been estimated that most of the time, the number of days without precipitation exceeded 330 days per year (i.e., less than 1 mm of rain per day) in the Maghreb between 1997 and 2008 (Cook 2012). Despite the decline in precipitation over the past three decades, the Maghreb has also witnessed some exceptional precipitation events that have triggered severe natural disasters (Daoud 2013).

In the Maghreb, the average annual precipitation is estimated to be around 357 km^3, equivalent to a water depth of 110 mm. This modest amount of precipitation can be attributed to the high level of aridity, which varies based on the different climatic zones present in the region. Approximately 88% of these precipitation volumes are absorbed by actual evapotranspiration, playing a significant role. These precipitations, primarily concentrated during the cold season (accounting for 80% of the total), exert their influence mainly along the coastal areas of Algeria and Tunisia, as well as the Moroccan Atlas Mountains.

In addition, multiannual drought periods represent a major challenge for the economies of the Maghreb countries. These droughts are becoming increasingly frequent and prolonged, as Stockton (1988) and Mutin (2011) point out. Surface water resources, which play a crucial role, are particularly dependent on rainfull fluctuations. In 2020, the average availability of surface water in this area was around 44 billion m^3/year, of which more than 65% was in Morocco.

1.3 *Increasingly High Utilization Indices*

The annual renewable water reserve allocated per person in the Maghreb region stood at 464 m^3 in 2020. This value notably remains below the theoretical scarcity threshold set by the World Health Organization (WHO 2015), which is set at of 1,000

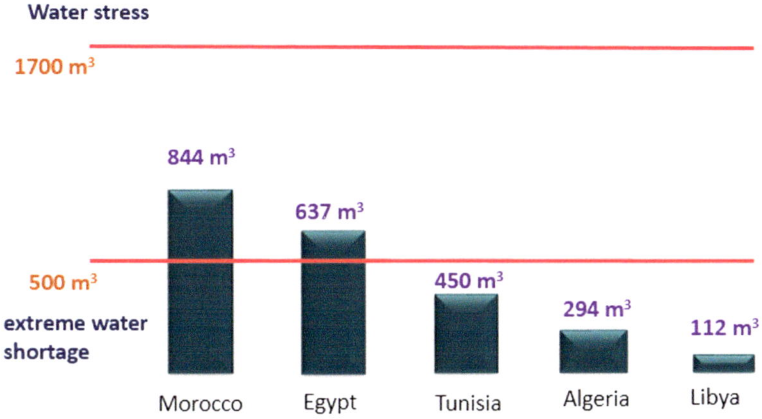

Fig. 1 Exploitable water resources (million m^3/capita/year). *Source* FAO—Food and agriculture organisation—AQUASTAT 2015

m^3 per person per year. This applies even to Morocco, which has the broadest access to surface water resources (Fig. 1).

Among the Maghreb countries. It is important to note that this allocation of water, although modest, can be misleading as it concerns renewable resources, not all of which are effectively usable. With population growth, this supply will fall to less than 290 m^3/capita/year by 2030. Algeria's already critical situation, at less than 290 m^3/capita/year, would worsen (in the absence of alternative non-conventional resources), as the supply would fall to around 180 m^3/capita/year.

1.4 Are Water-Related Risks Well Controlled and Mitigated ?

Despite the forecast increase in demand for water due to population and economic growth, alterations induced by climate change will play a predominant role in variations of water stress across the region. Climate change intensifies water stress through various mechanisms, including reduced precipitation and elevated temperatures (IPCC 2022). This trend increases evapotranspiration rates and irrigation requirements for crops (Verner 2012), alongside periods of intense heat (Lelieveld 2016). These factors will reinforce the water stress for surface waters in multiple regions and intensify precipitation irregularities.

Fragile nations, both politically and environmentally, will face specific water stress due to climatic disturbances. Simultaneously, climate change contributes to rising sea levels, thus increasing the risks of flooding and salinization of deltas and coastal aquifers. This saline water rise also impairs freshwater aquifers and river systems, particularly impacting coastal areas that overexploit groundwater as excessive extraction favors the intrusion of saltwater (Mabrouk 2013). Vulnerable populations are likely to be most affected by increased flood and drought risks, phenomena of particular concern in the Middle East and North Africa, where the most vulnerable populations are exposed to climate shocks (Hallegatte 2016).

Floods dominate observed natural disasters in this region, but the exacerbation and intensification of drought periods are becoming inevitable with climate evolution. Recent droughts far exceed the natural variability of the past millennia (Cook 2016), increasing concerns about their exacerbation due to climate change. Water scarcity poses difficult choices in the Middle East and North Africa region, where water, energy, and food intersect. The interactions between these complex sectors create challenging trade-offs that can have unforeseen consequences. Indeed, water is essential for food production, while energy is crucial for water production (desalination or pumping groundwater). The League of Arab States emphasizes in its Strategic Framework for Sustainable Development the importance of multisectoral approaches to address complex resource management issues (Gelil 2014).

2 Adaptation Strategy

Before the impacts of climate change gained widespread media attention, the Maghreb nations had already set up some strategies to enhance water supply through hydraulic mobilization policies. This approach primarily relied on dams construction and well drilling. Over the past five decades, regulated water volumes, which were close to the extracted volumes, have significantly increased, surpassing 22.50 km^3. However, in the face of growing water needs, these quantities have proved insufficient.

In this context, the official reports of Maghreb countries presented under the "United Nations Framework Convention on Climate Change" provide a valuable insight into the strategic vision guiding policymakers in the region, namely Algeria, Morocco, and Tunisia. By actively participating in international forums and establishing "National Committees on Climate Change," officials and experts in these three maghreban's countries strive to identify key elements for reorienting their water management policies towards an approach more focused on rationalizing water resources.

This approach aims to enhance water use efficiency by adopting measures such as regulated water pricing, promoting water-efficient irrigation techniques, and developing unconventional water supply sources, This approach aims to improve the efficiency of water use by adopting measures such as regulated water pricing, the promotion of water-efficient irrigation techniques and the development of non-conventional water supplies, but before listing a few strategies it is essential to give a vision of climate change mitigation.

2.1 Climate Change Mitigation Strategies

There are two broad categories of water mitigation strategies: technology-based and nature-based. In order to meet many of the global water challenges and deliver additional benefits in all areas of sustainable development, nature-based solutions (NBS) are an essential way to go beyond business as usual. NBS use or reproduce natural processes to increase water availability: soil moisture retention or groundwater recharge, improve water quality (for example, natural or artificial wetlands) and therefore reduce the risks of water-related disasters and climate change.

Ecosystems function as carbon sinks that absorb greenhouse gas emissions in SfN-based mitigation approaches. The preservation or restoration of wetlands, the reforestation of mangroves and the preservation of natural flood plains near rivers are just a few examples. Although peatlands, which are formed by peat soil and the wetland habitat that develops on its surface, account for only around 3% of the Earth's surface, they store at least twice as much carbon as all the Earth's forests. These interrelated hydrological and terrestrial and terrestrial ecosystems are an important resource for mitigating unexplored climate change mitigation resource.

Technological solutions to mitigate climate change generally require investment in reducing emissions from water infrastructure, such as drinking water supply, wastewater and storm water treatment, water pumping for agriculture and other uses. There are several strategies for reducing the risks associated with water and sanitation that need to be taken into account when planning and managing water abstraction, distribution and treatment operations in this context:

Increased energy efficiency: Energy savings of 10–30% can be achieved in water supply and wastewater treatment by installing energy-efficient pumps tailored to the specific characteristics of the network.

Renewable energy production and energy recovery: hereby reducing demand for fossil fuels.

Production of electricity using turbines installed along water supply and wastewater treatment networks treatment networks.

2.2 Widespread Water-Efficient Irrigation

With decreasing water availability, competition among various water uses becomes even more intense, especially as irrigation needs increase. These needs dominate in all countries; however, Algeria distinguishes itself by presenting a relatively moderate proportion (63%), in contrast to the rates observed in Morocco (86%) and Tunisia (80%). To alleviate pressure from agriculture on water reserves, irrigation methods known for their water consumption efficiency are gaining adoption.

2.3 Improving Irrigation Performance

Irrigated cultivated areas in the Maghreb region are estimated at around 12% (FAO 2017) of the total agricultural area used. Official data presents approximations due to the unpredictable nature of precipitation and, consequently, water availability. Nearly 56% of irrigated lands in the Maghreb are concentrated in Morocco. In contrast, Algeria and Tunisia have two and three times fewer irrigated lands than Morocco, respectively.

In the Maghreb region, areas with abundant water and those intensively used are at a certain geographical distance. To address this situation, substantial water transfers are undertaken, but these transfers result in significant water resource loss. This phenomenon is also observed in usage areas; where between 50 and 60% (Jihad 2012) of water is lost due to the lack of maintenance of supply channels and the use of gravity irrigation methods, especially in Morocco.

The Maghreb countries are confronted with the need to expand their agricultural lands through irrigation to ensure food security. By 2030, these areas could increase to 3.5 million hectares (FAO 2017), representing an approximately 30% increase from the current situation. However, this expansion is uncertain in Morocco and Tunisia due to concerns about water supply, given that the majority of their water resources are already utilized. In Algeria, technical and financial challenges limit the capacity to increase regulated water supply despite efforts to manage water flows.

Maghreb countries are implementing projects to promote more efficient irrigation methods, aiming to cover around 80% of irrigated lands by 2030 (Jihad 2012). These methods allow for increased agricultural production while limiting water consumption and soil erosion, with gains ranging from 40 to 60% compared to traditional irrigation.

2.4 Wastewater: A Valuable Resource

Currently estimated at around 2,330 hm^3/year for the three countries, wastewater discharges present approximate data, partly due to the frequent inclusion of rainwater from roofs and streets. The Maghreb region housed about 200 sewage treatment plants (35% in Morocco, 50% in Algeria, and 15% in Tunisia), with a theoretical annual treatment capacity of 1,125 hm^3. However, the amount of treated wastewater barely reaches 550 hm^3/year. Treatment rates range from 14% in Morocco to 39% in Tunisia for discharged waters. Treatment methods are generally secondary or even primary, suitable for uses like golf course irrigation or irrigating fruit trees, vineyards, and fodder crops. Nevertheless, these treatment levels prove insufficient when treated effluents are intended for irrigation of some vegetable crops or discharge into rivers. Tunisia stands out in terms of reclaimed water reuse, with a rate of 30% (Jihad 2012).

Significant amounts of untreated wastewater are discharged into the environment, leading to water quality deterioration. Sewage treatment plants, when present, are often ineffective due to insufficient maintenance. In Morocco, of about 70 existing plants, fewer than half operate regularly. Deteriorating water quality leads to environmental damage, estimated at about 2.7% of GDP for Tunisia, and around 4.4 and 4.9% of GDP for Morocco and Algeria, respectively. This plays a significant role in degrading water resources.

2.5 What Socio-economic Adaptation Ensures Equitable Access to Water Services?

Until the 1990s, water tariffs in Maghreb countries were low, contributing to significant losses in aging networks, estimated at half of the produced resources in 1995. These losses resulted from inadequate pricing that favored patronage and lack of maintenance. Since the late 1990s, improvements have been observed in Morocco and Tunisia due to regular tariff revisions, widespread meter usage, and abandonment of flat-rate billing in certain cities. In Morocco, the potable water recovery rate exceeds 99%, and this economic approach has led to reduced water consumption, particularly among heavy users. Tariff increases have also encouraged alternative practices. In Algeria, despite meter generalization, network efficiency remains low due to payment issues. In all countries, an economic approach has been adopted for potable water and sanitation management, with progressive tariffs promoting access and waste prevention.

2.6 Seawater: An Inexhaustible Resource

In the Maghreb region, desalination of seawater and brackish groundwater was first adopted to address potable water needs in isolated desert areas. Currently, production stands at about 0.8 hm^3/day, mainly from nearly twenty plants largely powered by conventional energy sources. In Morocco, the first seawater desalination facility was established in Tarfaya in 1976, followed by several other units, including the largest one in Laayoune, with a capacity of 66,000 m^3/day. Morocco currently has 12 existing stations with capacity118, 700 m^3/day and plans to increase this capacity to 396,200 m^3/d with this 7 stations under construction by 2030 (nouacer 2017).

On the southern coast of Tunisia, desalination of brackish water has primarily been developed since the 1980s, with a current capacity of over 100,000 m^3/day, projecting an increase of 5–6 times by 2030. However, the relatively high cost of seawater desalination limits its use to coastal desert areas. For other regions, desalination will become inevitable when conventional water resources are insufficient. This has already been the case for over a decade along the Algerian coastal strip, especially in cities and petrochemical complexes. Algeria, a regional leader in desalination, produces over 600,000 m^3/day in a dozen large plants and plans to reach 2.6 hm^3/day by 2030 (nouacer 2017).

3 Conclusion

In conclusion, climate change has emerged as a major global concern, directly or indirectly affecting all countries. To address this new challenge, a comprehensive approach is imperative. Southern nations, such as those in the Maghreb region, are particularly susceptible to these impacts. Each country must formulate its own responsive strategy to mitigate the effects of climate change.

However, the responsibility of Northern countries remains pivotal. They hold a critical role in providing scientific, technical, and financial support within the framework of international cooperation for sustainable development. Global collaboration is essential to collectively meet this challenge and safeguard the future of our planet. The Maghreb countries have already put in place strategic guidelines aimed at strengthening their research capacity on climate change and its effects. Although complex climate models are still difficult to assess on a fine and precise scale, they have the advantage of raising awareness of future risks among decision-makers in the Maghreb. However, few people are aware of these risks.

Effective adaptation to climate change will depend on the commitment and cooperation of the Maghreb countries in reducing greenhouse gas emissions, as well as on the adoption of "no-regrets" measures that will broaden the range of solutions to be developed in terms of water-saving irrigation techniques, wastewater reuse, seawater desalination, and so on. Delayed implementation of these measures risks compromising adaptation to climate change.

References

Agoumi A (2003) Vulnerability of Maghreb countries to change. International Institute for Sustainable Development, Geneva

Bank I (2018) Beyond shortages water security in the Middle East and North Africa. World Bank, Washinton

Barkhordarian AB (2013) The expectation of future precipitation change over the Mediterranean region is different from what we observe. Clim Dyn 225–244

Barkhordarian A, von Storch H, Zorita E (2012) Anthropogenic forcing is a plausible explanation for the observed surface specific humidity trends over the Mediterranean area. Geophys Res Lett 1–5

Cook BI (2016) Spatiotemporal drought variability in the Mediterranean over the Last 900 years. J Geophys Res Atmos 60–74

Cook K (2012) Impact of climate change on mid-twenty-firs century growing seasons in Africa. Clim Dyn 2937–2955

Daoud A (2013) Feedback on flooding in the agglomeration of Sfax (southern Tunisia) from 1982 to 2009: from prevention to territorialisation of risk. East's Geogr Rev 1–2

FAO (2017) Strategy 2018–21 FAO in North Africa. FAO Subregional Office for North Africa, Tunis

Gana A (2006) Water crisis in the Maghreb: resource scarcity or management problems? Géocarrefour 37–50

Garcia-Ruiz JL-M-S-M (2011) Mediterranean water resources in a global change scenario. Earth-Sci Rev 121–139

Gelil IA (2014) Proposal for an Arab strategic framework for sustainable development, 2015–2025. Economic and Social Commission for Arab High-Level Forum on Sustainable Development. AMMAn

Hallegatte SM-S (2016) Shock waves: managing the impacts of climate change on poverty. Word Bank, Washington

IPCC (2022) Climate change, global warming in the Maghreb. GIEC, CHARAM CHEIKH

Jaldi AS (2022) The climate and biodiversité situation in the Maghreb. charam cheikh: policy center for the new south.

Lelieveld JY (2016) Strongly increasing heat extremes in the Middle East and North Africa (MENA) in the 21st century. Clim Change 245–260

Mabrouk MB (2013) A review of seawater intrusion in the Nile delta groundwater system: the bas for assessing impacts due to climate changes and water resources development . Hydrol Earth Syst Sci 873–911

Thivet G (2009) Revising hydraulic strategies. In: CIHEAM and blue plan, Mediterranean 2009, rethinking rural development in the Mediterranean. Presses de Sciences Po, pp 65–99

Vizy EC (2017) Mid-twenty-first-century changes in extreme events over Northern and Tropical Africa. J Clim 5748–5767

Are Droughts Becoming More Severe or Longer in Morocco Based on the Standardized Precipitation Evapotranspiration Index for 1979–2022?

Sara Moutia, Mohamed Sinan, and Brahim Lekhlif

Abstract Morocco has been experiencing droughts in the last three decades. Furthermore, drought frequency and severity are expected to increase due to climate change. So, a performant drought monitoring is needed to contribute to the adaptation of more efficient measures to mitigate the opposite impacts. The monitoring of the drought is generally presented by the use of drought indices, which are based on different climatic and hydrological variables. In this study, the monthly. Standardized Precipitation Evapotranspiration Index (SPEI) data for the whole of Morocco is calculated Using ERA5 Reanalysis data relies on air temperature and precipitation. The SPEI is estimated by the time scale of 6 months depending on accumulated water balance for the six previous months. This work has two main objectives; firstly, an evaluation of ERA5 Reanalysis, to choose it as an alternative data of the ground observations using statistical metrics. Secondly, this work assesses the spatial and temporal features of drought obtained from the monthly SPEI. The results show that the drought irregularly appears in Morocco and that it is possible to distinguish drought years clearly using the SPEI. In addition, the Moroccan climate experienced a significant drying tendency with increasing spatial extent, concurrently drought events have also increased in terms of frequency, duration and intensity across Morocco examined by the comparison of the two 22-year periods 1979–2000 and 2001–2022.

Keywords Drought · SPEI · ERA5 reanalysis · Morocco · Climate change

S. Moutia (✉) · M. Sinan · B. Lekhlif
EHTP Ecole Hassania des Travaux Publics, Casablanca, Morocco
e-mail: moutia.sara@marocmeteo.ma

S. Moutia
DGM Direction Générale de la Météorologie, Casablanca, Morocco

N. Rebai et al. (eds.), *GIS, Applied Computing and Data Science for Water Management*, Lecture Notes in Geoinformation and Cartography,
https://doi.org/10.1007/978-3-031-63038-5_5

1 Introduction

Water plays an essential role in life, and in Morocco, a country full of geographical and climatic diversity, drought is a pervasive challenge that has profound consequences on daily life, agriculture, the economy and the environment. Throughout the past decades, Morocco has endured devastating droughts that have profoundly impacted its history, while our planet experiences unprecedented climate changes (Moutia et al. 2021). Located in the North-West of Africa, Morocco has suffered severe droughts during the recent decades, which caused large economic losses. Some governmental drought declarations were made in 1992–1995, 1998–2001, 2005, 2007, and 2015–2016. As an example, the 1994–1995 drought caused an estimated 7.6% Gross Domestic Product (GDP) loss. The 1999 drought cost an estimated US$900 million and touched more than one million hectares of farmland (Verner et al. 2018). Additionally, Morocco experiences high spatial and temporal fluctuations in rainfall, leading to variable water availability that significantly impacts the agriculture. Therefore, to lessen the adverse consequences of this natural hazard, effective drought monitoring and early warning systems are required.

Many other studies have attempted to provide spatial and temporal information on drought over Morocco (El Khatri and El Hairech 2014; Benassi 2008). What makes this work special is its basis on ERA5 reanalysis monthly data. These grided Reanalysis products address the limitations of ground observations, which can be spatially or temporally discontinue. However, their accuracy has to be evaluated and compared with ground 2m temperature and precipitation observations. The two statistical performance techniques Pearson correlation coefficient and Bias were used to evaluate the ERA5 Reanalysis with the corresponding ground observation data at twenty independent weather stations.

Drought can be generally defined as water deficits during a specific period, but it steals difficult to give an universal definition to quantify drought, this makes drought monitoring more difficult (Moutia et al. 2021). Based on several meteorological and hydrological factors, drought indices are used to track and evaluate the severity, length, and geographic extent of the drought(Sheffield and Pan 2015), the Standardized Precipitation Evapotranspiration Index (SPEI) considers in addition to precipitation the effect of evapotranspiration by incorporating the air temperature data but also has the multitemporal characteristic of the SPEI, which makes it a better alternative to the other indices (Vicente-Serrano et al. 2010; Huopo and Jianqi 2015). In this study, the SPEI is calculated based on the difference between total monthly precipitation and monthly potential evapotranspiration calculated using the Thornthwaite (1948) approach. It is based on the monthly 2-meter mean temperature.

In the context of the Moroccan climate, which has distinct wet and cool season Nov-Apr and dry and warm seasons Mai-Oct, the SPEI at the six-month time scale is chosen to capture these seasonal variations effectively. The SPEI helps assess drought conditions and water availability over these critical seasonal periods, providing valuable information for understanding and managing water resources and agricultural practices. The goals of this study were to provide information on the interannual variability of Standardized Precipitation Evapotranspiration Index characteristics, particularly, spatial extent and severity over the whole country, further a comparison between the two 22 years' periods 1979–2000 and 2001–2022 in terms of Drought frequency and duration and intensity across Morocco.

2 Study Area

Morocco is a country in the northwest of Africa and located at the level of the subtropical zone. One of the main characteristics of Morocco's climate is its great diversity, ranging from the temperate and humid type to the desert and dry type. Indeed, the Moroccan climate alternates between hot and cool periods depending on maritime effects from the Atlantic Ocean and Mediterranean and also Saharan influences from the Southeast and the South of the country (Born et al. 2009).

Based on interpolated data from the General Directorate of Meteorology (DGM), which provides a network of meteorological stations with high-quality measurements, Fig. 1 displays the map of annual precipitation normals during the base period 1981–2010. Morocco's rainfall patterns are visible; on average, the north-east receives more precipitation than the other regions of the country. The mountain ranges depicted in blue have annual rainfall totals over 600 mm. In addition, the average annual rainfall drops from north to south and from west to east. The map also displays the 20 weather stations, which span the majority of the nation, that were used in this study to evaluate the ERA5 Reanalysis.

Based on interpolated data from the General Directorate of Meteorology (DGM), which provides a network of meteorological stations with high-quality measurements, Fig. 1 displays the map of annual precipitation normals during the base period 1981–2010. Morocco's rainfall patterns are visible; on average, the north-east receives more precipitation than the other regions of the country. The mountain ranges depicted in blue have annual rainfall totals over 600 mm. In addition, the average annual rainfall drops from north to south and from west to east. The map also displays the 20 weather stations, which span the majority of the nation, that were used in this study to evaluate the ERA5 Reanalysis.

The climate of the country is characterized by two seasons, a warm dry season May to September and a wet and cold season October to April. Morocco is dominated by mountain ranges: Rif, Middle Atlas, High Atlas (3000 to 4000 m), and the Anti-Atlas, which subdivides it into two parts: a part in the west and north of the reliefs, which is influenced by the North Atlantic perturbations and an eastern and anti-Atlas part where the influences of the south-east Sahara dominate (Moutia et al. 2021; Born et al. 2009).

The length of the season presents slight spatial variations, particularly between the north and the south of the country. However, in most of the country, the months of October to March represent the wettest period, and the winter months (December, January, February) receive the most precipitation. The summer months have low rainfall and are generally marked by stormy precipitation (Xoplaki 2002).

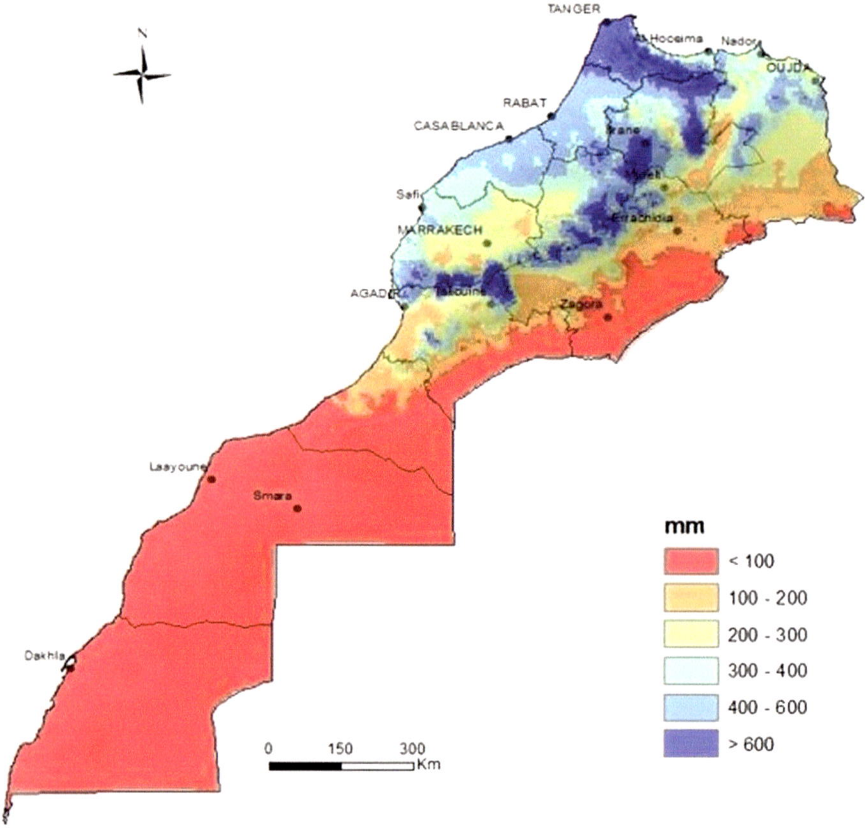

Fig. 1 Annual precipitation normals across Morocco derived from interpolated ground observation data from the General Directorate of Meteorology (DGM) for the years 1981–2010

Figure 2 illustrates the Walter–Lieth diagram of morocco. It shows the annual cycle of monthly rainfall sums (blue line, right axis) and monthly means of temperature (red line, left axis). These values are assumed by the temperature and precipitation averages of 20 weather stations dispersed throughout the country and covering the period 1981 to 2010. We can devise the months of the year into two major seasons. A wet and cool season from November to April and a warm and dry season from May to October. Each season has 6 months considering October such as a changing month from the hot and dry season to the wet and cool season.

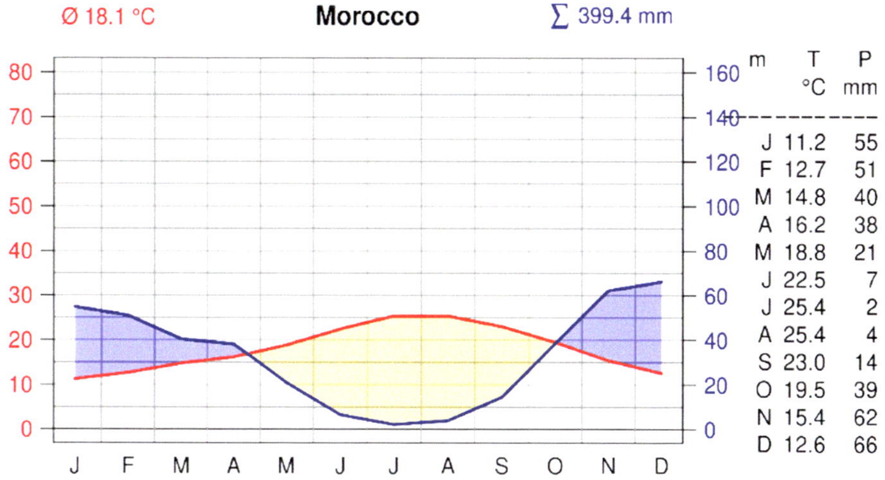

Fig. 2 The Walter–Lieth diagram for Morocco over the climate normal period 1981–2010

3 Data Used

3.1 *Weather Station*

The observed daily rainfall, maximum and minimum temperature data from 1979 to 2019 for 20 independent weather across the country were obtained from the General Directorate of Meteorology (referred to as DGM), The DGM has a network of meteorological observation stations covering the different climatic and geographical regions of Morocco. These different meteorological stations (or Provincial Meteorological Centers CPM) operate according to the standards in force of the World Meteorological Organization (WMO). In our case, we will use climate data from 20 stations (Fig. 1), which the missing data percentage does not exceed 1% in most cases. The quality and homogeneity of the data were checked (Driouech et al. 2021). Monthly data were calculated from the daily in situ observations for each weather station, these records were used for validation of ERA5 Reanalysis.

3.2 *ERA5 Reanalysis Data*

ERA5 is the most recent reanalysis product of the European Center for Medium-Range Weather Forecasts (ECMWF), it can be considered as one of the best reanalysis products in the world, with a spatial resolution of 0.25° (approximately 31 kilometers) and which is produced by combining ground-based observational data and satellite datasets and a numerical weather prediction model, which generates hourly weather records dating back to 1979 down to near real time (Hersbach et al. 2019). Hence, the mean advantage of ERA5 data, that it covers several decades and exists in areas that are poorly observed (Schepers et al. 2019). Reanalysis not only allow to increase the amount of data by integrating files from different sources, but also offer the possibility to improve their quality by cross-checking the data and testing it. A reanalysis, or a posteriori analysis, crosses data from direct or indirect observation and data from modelling. This unique combination makes it possible, compared to traditional observation, to obtain continuous spatial and temporal coverage on a global scale, and thus to be able to study the climate events in spaces with very low observation densities (Schepers et al. 2019).

4 Methods

4.1 Statistical Methods for ERA5 Reanalysis Evaluation

4.1.1 Pearson's Correlation Coefficient (R)

Pearson's correlation coefficient (r) is used to measure the well linear association between two variables. It measures the degree of conformity of two sources of data. Its value varies between 0 and 1, 1 indicates perfect correlation. It amounts the connection that exists between these two variables, we then speak of a linear adjustment. It is equal to −1 if a variable is a decreasing affine function of the other variable, or to 1 if one of the variables is an increasing affine function of the other variable. The degree of linear dependence between the two variables is revealed by intermediate values (Gleixner et al. 2020).

$$\frac{r = \mathrm{Cov}(\mathrm{Rn}, \mathrm{On})}{\sigma(\mathrm{Rn})\sigma(\mathrm{On})} \tag{1}$$

where On = Observations, Rn = the ERA5 Reanalysis data to evaluate.

Calculating the correlation coefficient (Eq. (1)) means calculating the ratio between the Covariance of the two variables On and Rn and the standard deviation of the variable Rn times the standard deviation of the variable On, the closer this coefficient is extreme values −1 and 1, the stronger the linear correlation between the variables; the two variables are said to be highly correlated.

4.1.2 Mean Error (ME)

The calculation of the mean error (ME) is one of the most used approaches to quantify the error during a comparison study. It is a parameter whose value makes it possible to give an order of the difference between two series of data, while its sign can indicate whether it is an underestimate or an overestimation (Xing et al. 2015).

$$\mathrm{ME} = \overline{\mathrm{R}} - \overline{\mathrm{O}} \tag{2}$$

where $\overline{\mathrm{R}}$ is the mean of the Reanalysis data and $\overline{\mathrm{O}}$ the mean of Observational measurements.

The mean error (ME) (Eq. (2)) is calculated as the difference between the means of the data sets. In our case, the mean error (ME) is calculated by subtracting the reanalysis data from the observed data. The closer this bias is to 0, the closer the reanalysis data is to the observed reality. The bias is positive if the reanalysis data overestimates the observed one, negative if it underestimates. The ME value of zero is the perfect score.

4.2 Standardized Precipitation Evapotranspiration Index (SPEI)

Drought indices are important tools for monitoring and assessing drought conditions. They help in understanding various characteristics of drought, including its intensity, duration, and spatial extent. These indices are based on different climatic and hydrological variables. They help decision-makers and researchers assess the severity and duration of drought events, which is vital for planning and mitigating their impacts on agriculture, water supply, and ecosystems. (Benassi 2008). Here are some key drought indices:

- Palmer Drought Severity Index (PDSI): One of the first drought indices to be created was the PDSI. It considers evaporation demand, moisture supply, and precipitation (Reichler and Kim 2008). However, a limitation of the PDSI is that it does not consider temporal scales, which means it may not capture drought conditions over longer time periods effectively (Palmer 1965).
- Standardized Precipitation Index (SPI): The SPI was developed to address the temporal scale limitations of the PDSI. It uses a probabilistic approach based only on precipitation data. By doing so, it provides a better understanding of drought conditions over different time scales (Yared et al. 2017).
- Standardized Precipitation Evapotranspiration Index (SPEI): The SPEI builds upon the SPI but improves it by incorporating air temperature data. This inclusion allows the SPEI to consider the effect of evapotranspiration, which is an essential factor in drought assessment (Vicente-Serrano et al. 2010). Additionally, the SPEI has a multitemporal characteristic, making it a better alternative to both the PDSI and SPI (Sheffield and Pan 2015).

SPEI is calculated using the 'SPEI' R package as the difference between the total monthly precipitation and monthly potential evapotranspiration (PET).

Table 1 Drought classification based on the SPEI

Drought category	SPEI value
No drought	$0 \leq \text{Index}$
Light drought	$-1 \leq \text{Index} < 0$
Moderate drought	$-1.5 \leq \text{index} < -1$
Severe drought	$-2 \leq \text{Index} < -1.5$
Extreme drought	$\text{Index} < -2$

Monthly PET was calculated by the Thornthwaite (1948) approach based on monthly 2m mean temperature. Additional details of the SPEI, are described in Sheffield et al. (2015). The six months' time scale was selected for this study, it depends on the accumulated water balance for the six previous months (Zhang et al. 2006). According to the six-month seasons that characterized the Moroccan climate, Nov–Apr as a wet and cool season and Mai-Oct as a dry and warm season.

Drought levels has been defined as five types (Vicente-Serrano et al. 2010) presented in Table 1.

4.3 Drought Related Indicators

4.3.1 Percentage Drought Area

Percentage drought area is the ratio of the number of drought pixels (where the SPEI < -1) to the total number of pixels which equals 1024. It displays the spatial extent of drought occurrence (Potopová et al. 2015; Meixiu et al. 2014).

4.3.2 Duration and Intensity of Drought Events

A drought event can be defined as a period in which the SPEI continuously reaches a value of -1.0 or less at least for one month (Li et al. 2012). Following the selection of a drought event, drought-related metrics such as intensity, duration, and frequency were computed. A drought occurrence lasts for as many months as it does. The duration of a drought event equals the number of its months (Chunping et al. 2015). The intensity is the absolute value of the sum of all SPEI values during a drought even divided by its duration.

5 Results and Discussion

5.1 *Evaluation of ERA5 Reanalysis*

In this study, twenty weather stations were investigated to evaluate the ERA5 Reanalysis to choose it as a good alternative for the spatial and temporal assessment of drought in Morocco on a monthly time scale for the period from 1979 to 2019. Similar to the observational datasets, the ERA5 Reanalysis of precipitation and temperature were aggregated to the monthly SPEI of 6-months' time scale. For comparison purposes, statistical techniques, specifically, Pearson correlation coefficient (r) and Mean Error (ME) are calculated.

Figure 3 shows the map of correlation statistical metrics obtained for weather stations against the ERA5 Reanalysis in terms of the monthly SPEI. In general, the result shows a good agreement between the two sources of data, with a higher correlation coefficient value above or near **0.8** for all weather stations, except Essaouira, Hoceima, and Laayoune where the correlation coefficient is near **0.7**. Relatively, a higher correlation coefficient with **0.915** as a value was obtained in Meknes. The lowest correlation coefficient **0.69** was obtained in Laayoune. This shows that the two sources of data are well correlated in terms of the monthly values of SPEI.

In our case, Mean Error (ME) is calculated by subtracting the average of ERA5 Reanalysis from the average of DGM data for the twenty weather stations. Figure 4 shows that the ME is relatively closer to 0 for ten stations and less than 0.2 for the rest. Hence, the reanalysis data little underestimate the drought in half of the total number of weather stations.

5.2 *Spatial and Temporal Variation of Drought*

Our goal was to provide basic interannual spatial and temporal information for drought assessment in Morocco based on the SPEI values. Usually, the SPEI values at the 6-month scale for the month of April indicate the state of the cool and wet season from November to April. Respectively, the October SPEI indicates the water deficit caused by drought during the hot and dry season from May to October. Then, to get an idea of the drought in Morocco, we averaged the SPEI values over the entire Moroccan territory (1024 grid points). Then we calculated the percentage of the drought area according to 3 levels: moderate drought with a SPEI between -1.5 and -1, severe drought with a SPEI between -2 and -1.5 and extreme drought with a SPEI < -2 (See Table 1).

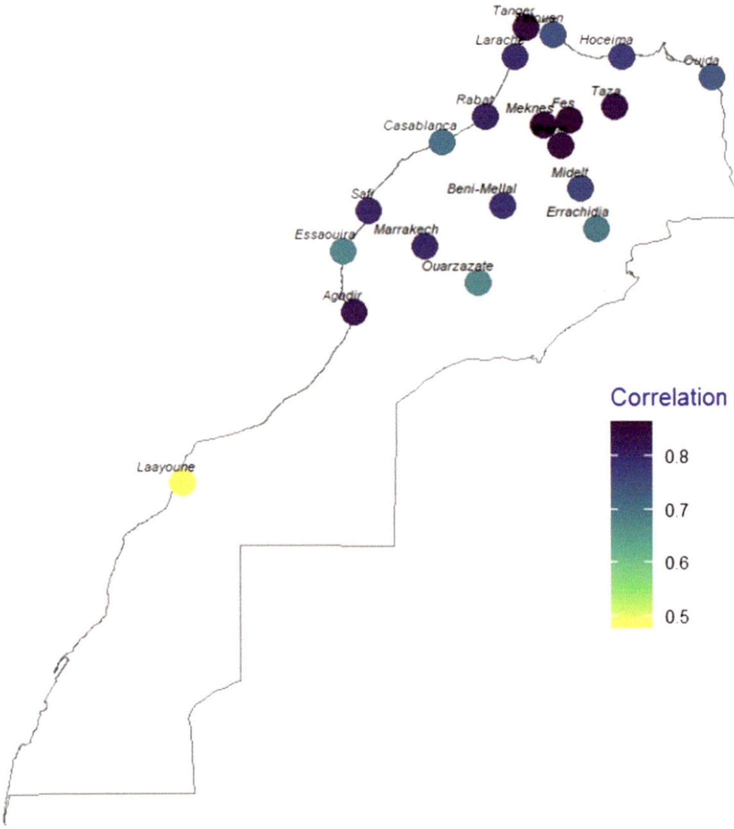

Fig. 3 Correlation map of ERA5 reanalysis versus observations

5.2.1 November–April Season

Figure 5 shows the interannual series of SPEI of whole Morocco during the years 1979–2022 for the wet and cool season (November-April), it illustrates that drought appears irregularly in Morocco, also, it is possible to easily distinguish wet and dry years using the 6-month SPEI. In addition, a careful reading of the graph indicates that the number of seasons November-April from 1979 to 2022 that experienced droughts is 23, while seasons that were wet is 16. Prior to the year 2000, most overall SPEI values are positive (13 values in blue), and just 8 values that have less than 0 marking drought (in red); only the value of the season corresponding to the year 1986–1987 is close to −1.0, thus close to indicating a moderate drought; however, just after 2000, we notice that the number of negative values indicating drought (14 values) is more than the double of the positive values (6 values); 4 of them are below −1.0 for the November-April season. Namely, for the years 2000–2001, 2007–2008,

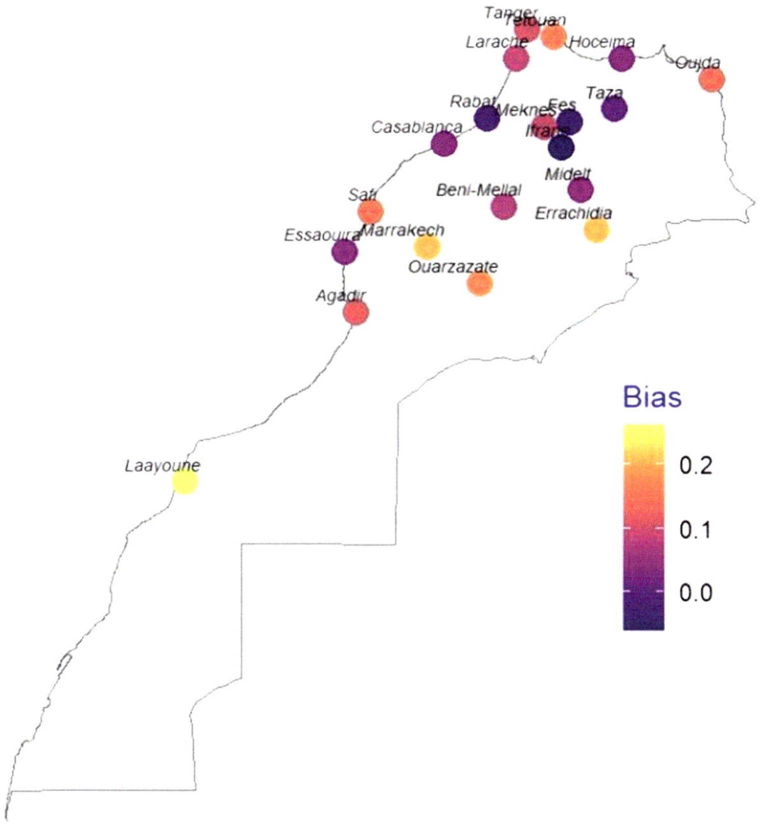

Fig. 4 Mean error map (bias) of ERA5 reanalysis versus observations

2015–2016 and 2019–2020. The most severe drought occurred in November–April 2015/2016 with an SPEI value exceeding—1.5.

Figure 6 shows the interannual variation of the spatial extent of drought to see the spatial distribution of different degrees of drought, because an averaged value over all Morocco even if it is useful it can hide behind it a set of important results.

The temporal variation of the percentage of drought area in Morocco is consistent with the previous results concerning observed national SPEI values. The Nov-Apr seasons that have experienced more than 50% spatial extent of drought are those having an SPEI value less than −1 indicated before, namely 1986–1987, 2000–2001, 2007–2008, 2015–2016 and 2019–2020, also, the two seasons 2009–2010 and 2020–2021. Especially, the 2015–2016 season has experienced the greatest extent of drought of 74%, followed by 2007/2008 with an extent of 70%. In fact, even the November-April 2009–2010 season has experienced a light drought in terms of SPEI value, it experienced the highest percentage of extreme drought area of 17% in a total drought area of 52%.

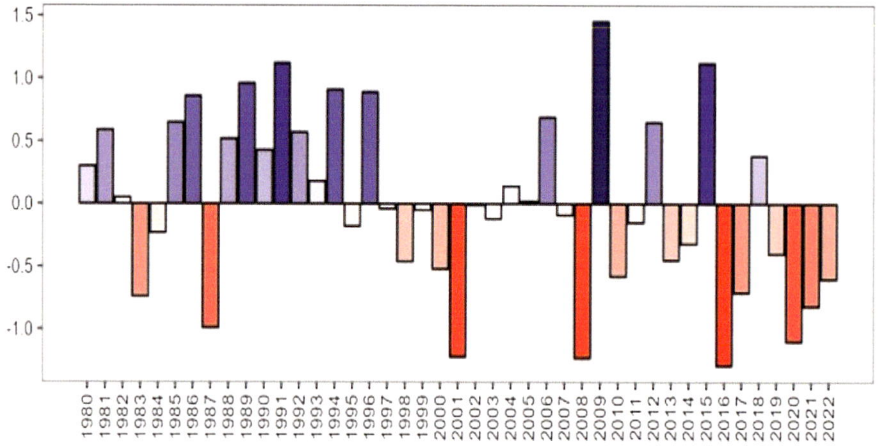

Fig. 5 Annual variation of SPEI values for the November–April season

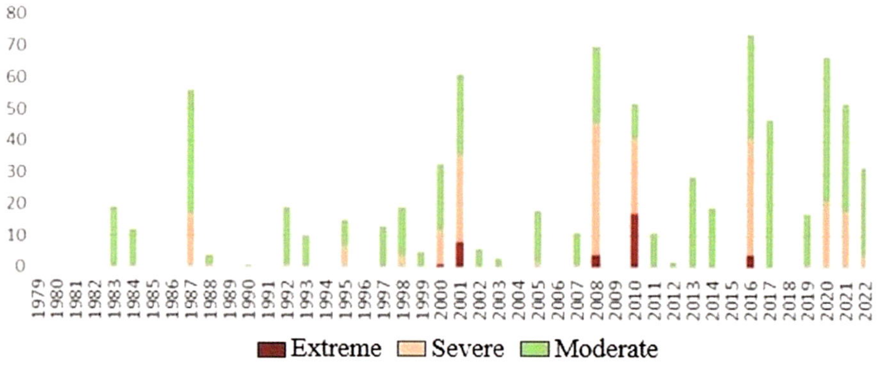

Fig. 6 Interannual variability of the drought area for the three drought levels: moderate, severe, and extreme

As noticed before, the areas influenced by extreme drought (SPEI < −2) are zero before the year 2000, also, the extent area of drought has increased noticeably after this transitional year of 2000.

5.2.2 Mai-October Season

The interannual series of SPEI for the whole of Morocco during the years 1979–2022 for the May-October season reveals that before the year 2000, most of the global values of SPEI are positive (15 values), just 5 values are lower than 0 but also above −0.5 indicating a light drought. However, just after 2000, the situation was reversed, almost all SPEI values are negative, 4 values of them even reach −1.5 so a severe drought for the May-October season, namely, the seasons corresponding to the years, 2017, 2020, 2021 and 2022. The extreme drought has occurred in 2017 and 2022 with an SPEI value close to −2. These extreme SPEI values that marked these years are mainly due to the high temperatures that characterized the summer seasons of these years (Fig. 7).

Figure 8 showing the temporal variation of the percentage of drought area in Morocco during the period 1979–2022 for the different degrees of drought confirm that the areas influenced by drought are less than 15% before 2001 during the May-October seasons. While during the 21st century the extreme drought appears and the percentage of spatial extent of drought has seen a remarkable increase, it has even exceeded 80% for the seasons Mai-Oct of 2017, 2020, 2021 and 2022. In addition,

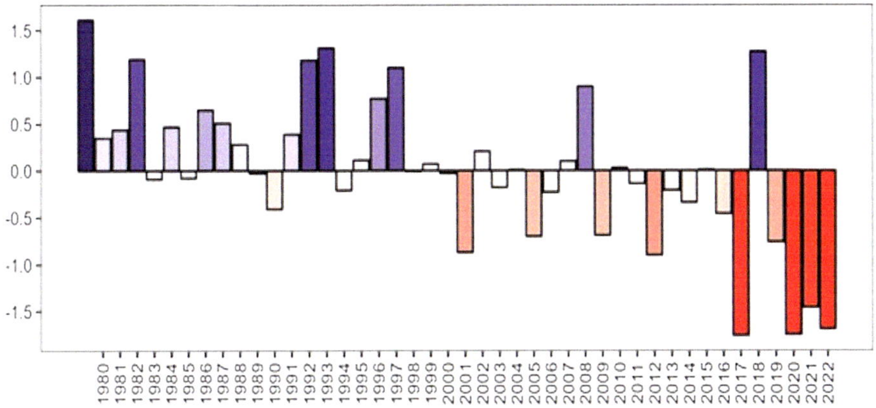

Fig. 7 Annual variation of SPEI values for the May–October season

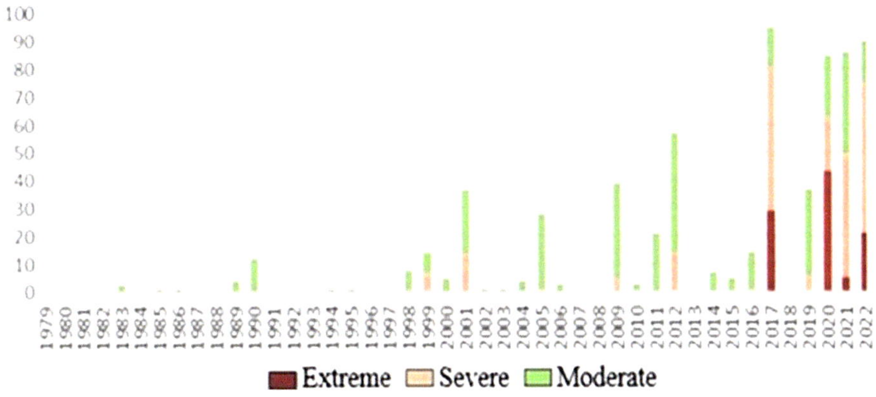

Fig. 8 Interannual variability of the drought area for three levels of drought: moderate, severe, and extreme for the May–October season

Table 2 Duration, intensity and frequency of drought events identified for the periods 1979–2000 and 2001–2022 in Morocco

Drought indicators	1979–2000	2001–2022
Number of drought events	9	15
Average duration (months)	2	4
Maximum duration (months)	6	11
Average intensity	1.27	1.48
Maximum intensity	1.48	1.87

the extreme drought (SPEI ≤ -2) has the maximum extent in 2022 with 42% of area followed by the year 2017 by a percentage of 29%.

5.3 *Drought Related Indicators*

The previous results reveal that the year 2000 was a transitional year which marked the beginning of a period more affected by drought, so we divided the study period 1979–2022 into two periods of 22 years, 1979–2000 and 2001–2022 to compare the drought indicators of these two periods. We calculated the average duration, the maximum duration, the average intensity and the maximum intensity and the frequency for each grid point for each period, then we calculated the averages over the whole of Morocco, the following table illustrates the Drought indicators, including drought frequency in % which expresses the occurrence of drought, duration and intensity for 1979–2000 and 2001–2022 (Table 2).

The occurrence of drought during the period 2001–2022 increased from 9 to 15 drought episodes compared to the period1979–2000. The average duration of drought events calculated from the SPEI for 2001–2022 increased by two to four months compared to the period 1979–2000, at the same time, the national maximum duration of drought events increased from 6 to 11 months. The average drought intensity also increased from 1.27 to 1.48, and from 1.48 to 1.87 as an intensity.

5.4 *Drought Trends*

This section presents the trend map of SPEI data calculated by ERA5 reanalysis at a six-month time step, which is used to identify the degree of drought based on both rainfall and evapotranspiration information in a given region. The linear trend of the SPEI drought index is calculated at each grid point, using the Mann–Kendall method, a non-parametric method, in order to obtain more reliable estimates of the magnitude and statistical significance of the trend (Driouech et al. 2021).

The map shows five illustrative degrees of the drought trend, each degree corresponds to a range of well-defined values. Ranging from a very weak trend to a very strong trend. These values are estimated by the values of the 20, 40, 60, and 80%

percentiles, correspond to the values −0.33, −0.26, −0.19, and −0.12 respectively. So, a very weak trend corresponds to values of trends greater than −0.12, a weak trend between −0.12 and −0.19, a medium trend between −0.19 and −0.26, a strong trend between −0.26 and −0.33, and a very strong trend to values below −0.33.

Figure 9, of trends from 1979 to 2022 informs us that the trend is towards the decline of the SPEI index equivalent to worsening drought conditions, with important and significant values. The trends calculated by ten years are very significant with p-values below 0.05, almost 80% of the values are below 3.10^{-9}.

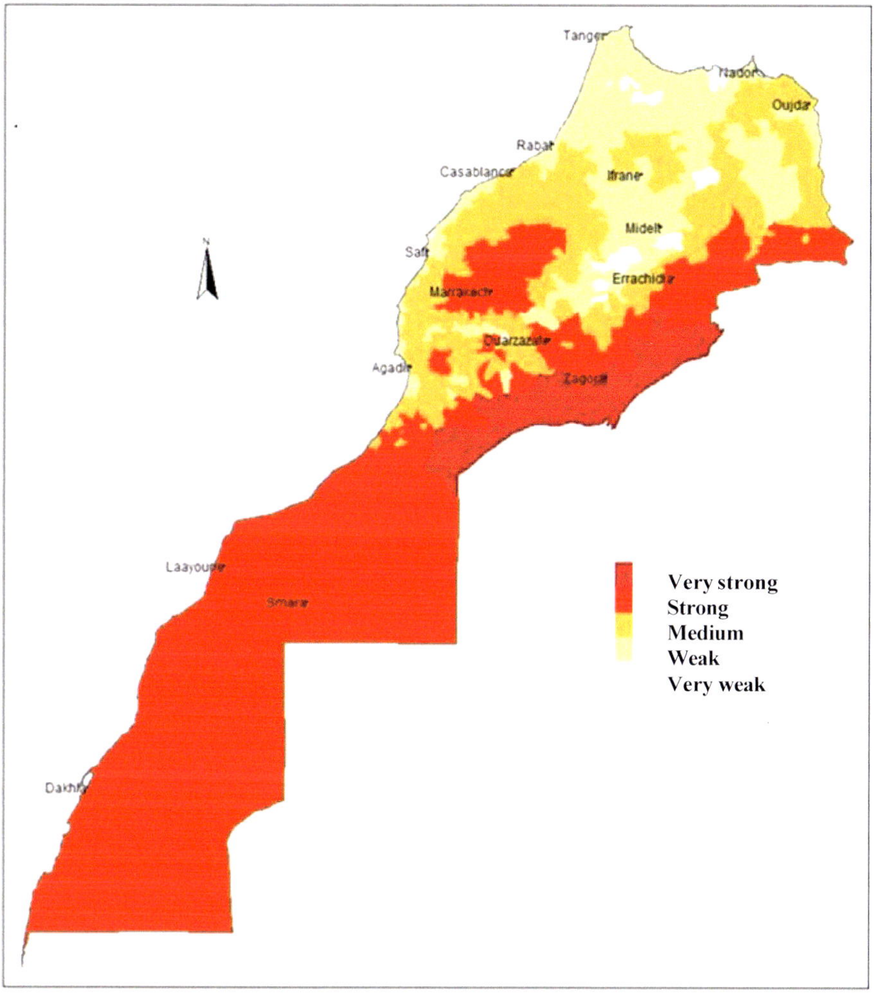

Fig. 9 Significant drought trends per ten years map in Morocco based on SPEI for the period 1979–2022

Starting with the least dominant case of the very weak drought trends, which corresponds to trend values between −0.04 and −0.12 per 10 years According to the SPEI trends map, they are present in the mountainous regions of the Rif and Middle and High Atlas. For the weak trends between (−0.12 and −0.19) and medium trends between (−0.19 and −0.26), they are present almost throughout northern Morocco, except the regions between Khouribga and Marrakech where the drought trends are strong (values between −0.26 and −0.33/10 years). In addition, the strong trends are dominant throughout the southern half of the kingdom and some regions of the southwest. Regarding the case of a very strong drought trends (values between −0.4 and −0.33/10 years), they are concentrating in the regions of the extreme south-west of the country. It can be seen that in the northern half of the country the trends become stronger with the continentality going from the Mediterranean and Atlantic coasts to the interior of the country. However, the higher regions of the High and Middle Atlas have very weak drought trends.

6 Summary and Conclusions

The availability of ERA5 Reanalysis data at spatial and temporal scales is beneficial to fill the ground observations gaps. However, their accuracy evaluation is essential for any application including drought assessment. In the present study, we have examined ERA5 reanalysis compared to ground observations, two statistical metrics were investigated at a monthly time scale by comparing the values of SPEI calculated from ERA5 Reanalysis with those from in situ observations from 20 independent weather stations. After analyzing the results, the performance of the ERA5 Reanalysis in terms of monthly SPEI was reasonably good in Morocco and can be used as a favorable alternative to ground observations with better spatial and temporal availability in the study area.

In brief, the results revealed by the SPEI over the period 1979–2022 show that drought appears irregularly in Morocco and that it is possible to clearly distinguish the years of drought using the SPEI. Morocco has experienced a significant trend towards drier conditions, the results showed a decrease in the slope of average SPEI values of 0.3 per decade during the years 1979–2022 thus a worsening of drought.

The same applies to the spatial extent of drought calculated from the SPEI, which shows a significant rate of increase of 7% per decade during the rainy season from November to April. In addition, a comparison of the characteristics of drought between the periods 1979–2000 and 2001–2022 revealed an increase of 9 to 15 drought episodes. The average and maximum duration of drought events increased by two months and five months, respectively, while the average and maximum drought intensity increased by 0.21 and 0.31. Hence, more frequent and longer drought episodes, more severe drought and larger spatial extent.

In addition, the Mann–Kendall method is employed to calculate linear trends of the Standardized Precipitation Evapotranspiration Index drought trends in Morocco using the ERA5 reanalysis data from 1979 to 2022 at each grid point, revealing a significant decline in the SPEI index, indicating worsening drought conditions. The northern half of Morocco except the regions between Marrakech and Khouribga, generally exhibits medium to weak drought trends, particularly in mountainous regions with very weak drought trends. The southern half and southwest regions show dominant strong and very strong drought trends.

Generally, the SPEI is suitable for application examining drought assessment. In addition, further studies are needed to deepen the understanding of drought variation in Morocco to monitor its risk.

References

Benassi M (2008) Drought and climate change in Morocco. Analyze of precipitation field and water supply. Opt Méditerranéennes 80:83–87

Born K, Christoph M, Fink AH, Knippertz P, Paeth H, Speth P (2009) Moroccan climate in the present and future: combined view from observational data and regional climate scenarios. In: Zereini F, Hötzl H (eds) Climatic changes and water resources in the Middle East and North Africa. Environmental science and engineering (environmental science). Springer, Belin

Chunping T, Jianping Y, Man L (2015) Temporal-spatial variation of drought indicated by SPI and SPEI in Ningxia Hui Autonomous Region, China. ISSN 2073-4433

Driouech F, Stafi H, Khouakhi A, Moutia S, Badi W, ElRhaz K, Chehbouni A (2021) Recent observed country-wide climate trends in Morocco. Int J Climatol 41(Suppl. 1): E855–E874

El Khatri S, El Hairech T (2014) Drought conditions and management strategies in Morocco

Gleixner S, Demissie T, Diro GT (2020) Did era5 improve temperature and precipitation reanalysis over East Africa? Atmosphere. MDPI

Hersbach H, Bell B, Berrisford P, Horányi A, Sabater JM, Nicolas J, Radu R, Schepers D, Simmons A, Cornel Soci DD (2019) Global reanalysis: goodbye ERA- interim, hello ERA5. ECMWF Newsl 159:17–24

Huopo C, Jianqi S (2015) Changes in drought characteristics over china using the Standardized Precipitation Evapotranspiration Index. J Clim 28:5430

Li YJ, Zheng XD, Lu F, Ma J (2012) Analysis of drought evolvement characteristics based on Standardized Precipitation Index in the Huaihe River Basin. Proc Eng 28:434–437

McKee TB, Doesken NJ, Kleist J (1991) The relationship of drought frequency and duration to time scales. In: Proceedings of the eighth conference on applied climatology, Boston, MA, USA, 17–22 January 1991, pp 179–184

Meixiu Y, Qiongfang L, Hayes MJ, Svobodab DM, Richard R (2014) Heime are droughts becoming more frequent or severe in China based on the Standardized Precipitation Evapotranspiration Index 1951–2010? Int J Climatol 34:545–558

Moutia S, Sinan M, Lekhlif B (2021) Assessment of agricultural drought in Morocco based on a composite of the Vegetation Health Index (VHI) and Standardized Precipitation Evapotranspiration Index (SPEI). In: E3S web of conferences, vol 314, p 04003

Palmer WC (1965) Meteorological drought. Research paper, 45. https://www.ncdc.noaa.gov/temp-and-precip/drought/docs/palmer.pdf

Potopová V, Štěpánek P, Možný M, Türkott L, Soukup J (2015) Performance of the standardized precipitation evapotranspiration index at various lags for agricultural drought risk assessment in the Czech Republic. Agr Forest Meteorol 202:26–38

Reichler T, Kim J (2008) Uncertainties in the climate mean state of global observations, reanalyses, and the GFDL climate model. https://doi.org/10.1029/2007JD009278

Sheffield, Pan M, Wood EF (2015) Evaluation of the tropical rainfall measuring mission multi-satellite precipitation analysis (TMPA) for assessment of large-scale meteorological drought. Remote Sens Environ 159:181–193

Verner D, Treguer D, Redwood J, Christensen J, Mcdonnell R, Elbert YK, Belghazi S (2018) Climate variability, drought, and drought management in Morocco's agricultural sector. World Bank

Vicente-Serrano SM, Beguería S, López-Moreno JI (2010) A multiscalar drought index sensitive to global warming: the standardized precipitation evapotranspiration index. J Clim 23:1696–1718

Xing L, Binbin H, Xingwen Q, Zhanmang L, Xiaojing B (2015) Use of the Standardized Precipitation Evapotranspiration Index (SPEI) to characterize the drying trend in Southwest China from 1982–2012. Remote Sens 10917–10937. https://doi.org/10.3390/rs70810917

Xoplaki E (2002) Climate variability over the Mediterranean. PhD, Inaugural dissertation der Philosophisch-naturwissenschaftlichen Fakultät der Universität Bern

Yared B, Tsegaye T, Getachew BD, Andualem S, Yared B, Tsegaye T, Getachew BD, Andualem S (2017) Evaluation of satellite-based rainfall estimates and application to monitor meteorological drought for the Upper Blue Nile Basin, Ethiopia, Drought Mitigation Center Faculty Publications, University of Nebraska, Lincoln

Zhang Q, Liu CL, Xu CY, Xu YP, Jiang T (2006) Observed trends of annual maximum water level and streamflow during past 130 years in the Yangtze River Basin, China. J Hydrol 324:255–265

GIS and Numerical Modeling

Assessing the Future Climate Change Impacts on the Foum El Oued Groundwater Aquifer in Laâyoune, Morocco: A Numerical Modeling Approach

Abdelkader Larabi, Mohamed Jalal El Hamidi, and Mohamed Faouzi

Abstract Water resource management in arid regions requires the provision of decision support tools that consider the impact of climate change as a structural reality. In the Laâyoune region, located in the south of Morocco, the Foum El Oued aquifer serves as a significant fresh water reservoir in a coastal area and plays a vital role in supplying water to the city of Laâyoune and neighboring municipalities. This study focuses on understanding and preserving this essential water resource, which is extensively utilized for drinking water supply, agriculture, and industrial purposes. However, due to the combination of reduced natural replenishment from the infiltration of Oued floods and increased pumping activities, groundwater has been overexploited, leading to the degradation of water quality and a considerable decline in the piezometric level. Consequently, an imbalance has emerged at the natural saltwater interface, resulting in seawater intrusion into the freshwater to compensate for the created depression. To address this issue, the objective of this work is to develop a finite difference hydrodynamic and seawater model that considers the effects of future climate change on the Foum El Oued aquifer, based on available hydrogeological data. The simulation results reveal that by 2050, under the projected conditions, the interface toe would have advanced by a distance of 6.5 km. At this time, the concentration of 10 g/l will reach a distance of 6.2 km, and the highest concentration (35 g/l) will extend to 5.8 km. This GW model serves as a crucial decision support system for water resource management, particularly for the population of Laâyoune and the irrigation of the Foum El Oued agriculture area. By implementing effective management strategies, the coastal area can be protected from contamination, ensuring its sustainable use and preventing abandonment due to increasing salinity levels in irrigation and pumping wells.

A. Larabi · M. J. El Hamidi (✉) · M. Faouzi
Regional Water Center of Maghreb, LAMERN, EMI, Mohamed V University in Rabat, Rabat, Morocco
e-mail: elhamidi@emi.ac.ma

N. Rebai et al. (eds.), *GIS, Applied Computing and Data Science for Water Management*, Lecture Notes in Geoinformation and Cartography,
https://doi.org/10.1007/978-3-031-63038-5_6

Keywords Water resource management · Impact of climate change · Foum El Oued aquifer · Hydrodynamic and seawater model · Hydrogeological data · Water resource management

1 Introduction

The significant decline observed in the Foum El Oued aquifer underscores the necessity for a numerical tool capable of accurately simulating the hydrodynamic behavior and seawater intrusion in this aquifer system. Hence, the aim is to understand the hydrogeological characteristics and to formulate a comprehensive conceptual model and subsequently a hydrodynamic/transport model under both steady and transient conditions.

For this purpose, a steady-state flow model is accurately designed to produce the approximate magnitudes of the hydrodynamic parameters and flow directions. Based on previous studies and data collected from the Agence du Bassin Hydraulique Sakia El Hamra and Oued Eddahab (ABHSHOD) (ABH SHOD 1993, 2003, 2005, 2009, 2010; 2013a–c), this steady-state model is designed to incorporate the available knowledge of the hydrogeological setting. This approach will facilitate a comprehensive analysis of the data, serving the dual purpose of validating the information and enhancing comprehension of the mechanisms pivotal to the eventual realization of the conceptual model.

Both models are developed for this aquifer system using the MODFLOW code, which is integrated into the Visual MODFLOW 2015 software.

2 Hydrogeological Setting and Extension of the Model

This aquifer is located between the city of Laayoune to the east and the coastal area of the ocean to the west, covering a width of approximately 17 km and an area of around 350 km^2 (refer to Fig. 1a). The Foum El Oued aquifer basin is extended along the ocean coast of the sedimentary basin encompassing Laayoune, Boujdour, and Dakhla. This basin consists of various geological layers ranging in age from the Triassic to the recent Quaternary.

Within the geological sub-basin of Foum El Oued, the aquifers mainly comprises post-Upper Miocene formations characterized by complex combinations of detrital, biochemical, and chemical sediments, including sands, conglomerates, clays, marls, limestones, chalks, evaporites, and more. This aquifer manifests as a multi-layered system, with the primary horizons categorized into four principal units: (i) lower sands, (ii) dolomitic limestones and lower sandstones, (iii) upper sands, and (iv) the Plio-Quaternary alluvial deposits of the Saguia El Hamra wadi. The aquifer's thickness fluctuates in accordance with the depth of the impermeable bedrock, varying from a few meters to around 200 m.

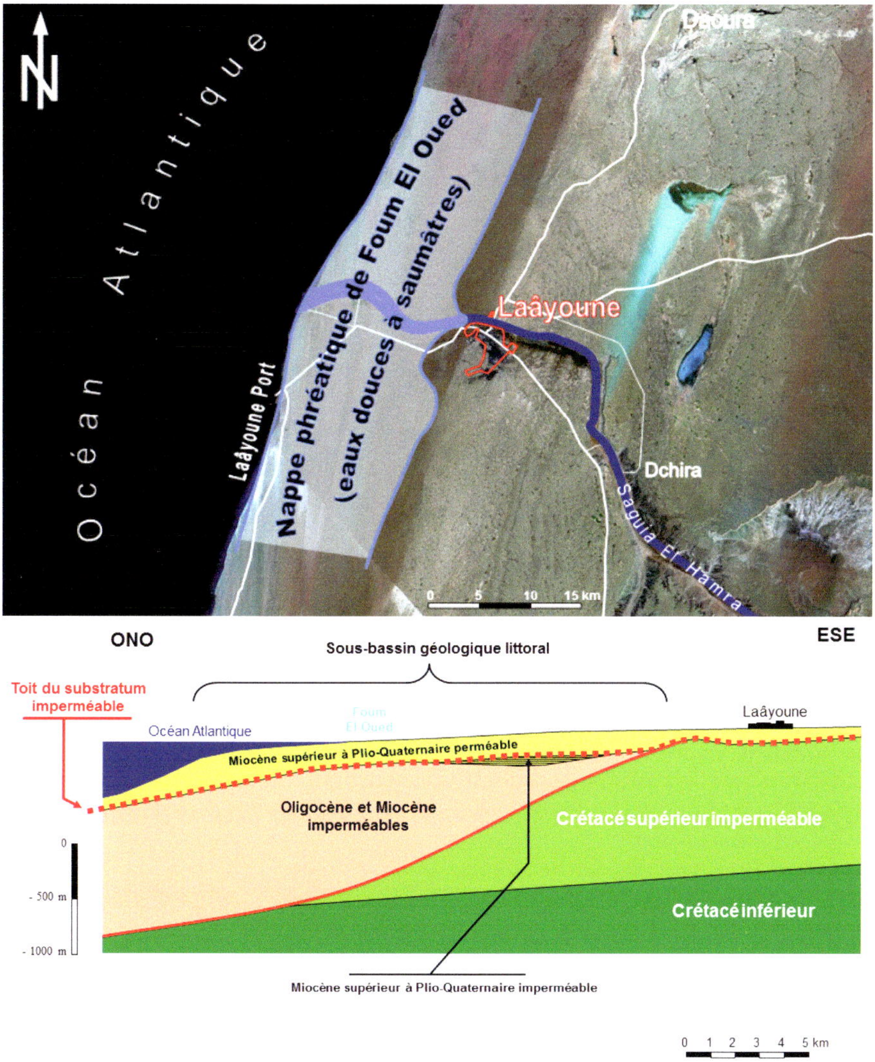

Fig. 1 **a** Aquifer location in proximity to the city of Laâyoune, **b** schematic cross-section of the Foum El Oued geological sub-basin

Beneath these formations lies an impervious bottom, primarily composed of clay-rich sediments from the Upper Cretaceous to Upper Miocene epochs. These sediments exhibit colors ranging from grey to green, blackish, or whitish and include clays and silty-sandy marls. The upper boundary of the aquifer generally resides within the upper sands and the Plio-Quaternary alluvium of Saguia El Hamra Wadi.

The delineations of the northern and southern limits of the Foum El Oued aquifer remain uncertain due to the lack of boreholes and further electrical soundings.

Nevertheless, the prevailing flow direction of the aquifer predominantly follows an east-to-west trajectory, ultimately converging towards the Atlantic Ocean.

3 Discretization of the Study Area

The study area was subdivided into a grid composed of squares/rectangles, each designed to accurately represent the aquifer system's geometry and structure.

A finer grid translates to enhanced accuracy in representing borehole locations and aquifer contours, consequently yielding highly dependable outcomes. For our specific numerical model, we opted for a spatial discretization interval of 250 m, resulting in a grid layout of 101 columns and 81 rows. This choice ensures an optimal balance between computational efficiency and accurate representation.

Based on the available hydrogeological conditions, boundary conditions are set up as described in Fig. 2a. The hydrodynamic parameters used, such as transmissivity values have been determined from pumping tests at the ONEE well field. The recharge resulting from floods in Saguia El Hamra Wadi is illustrated in Fig. 2b.

4 Steady-State Calibration and Sensitivity Analysis

The initial K-values were introduced in the model based on pumping tests, and modified during the calibration process in steady state regime. The map (Fig. 3a) illustrates the result of the model calibration for the distribution of K-values.

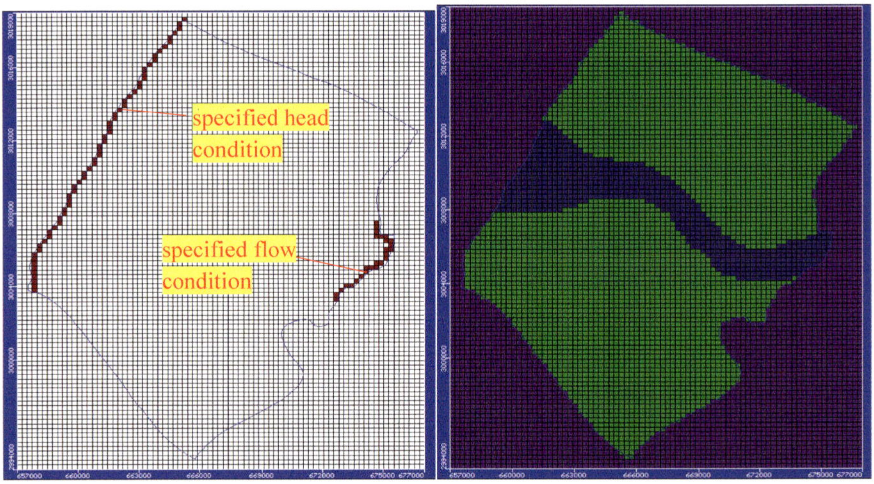

Fig. 2 **a** Mesh configuration and boundary conditions, **b** GW recharge map

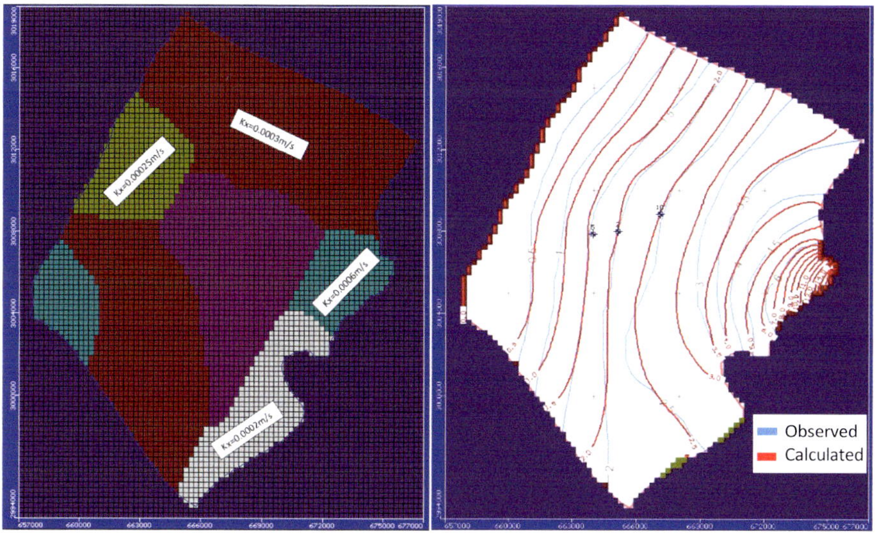

Fig. 3 **a** Calibrated K-values map, **b** comparison between measured and calculated piezometric data (1976 state)

The results demonstrate that the correlation coefficient in three piezometers, 2/120, 5/120, and 10/120, is notably high at 99.6%, indicating a strong agreement (Fig. 3b). Furthermore, it is noteworthy that the calculated balance for the year 1976 closely aligns with the observed balance of the same year. Following this phase, we conducted sensitivity tests to apprehend how the model responds to altered conditions such as multiplying calibrated K-values with corrective factors of 0.5 and 1.5. This results in changes of the water table's map, with an elevation of the piezometric level evident when K are multiplied by the corrective factor of 0.5. While, a reduction in the piezometric level is noted when K-values are multiplied by 1.5.

Upon completing the steady-state model design, we progressed to develop the transient state model based on available data from 1976 to 2019, including the storage coefficient parameters of the aquifer. This model was also calibrated to accurately reproduce the annual piezometric records at selected measurement points in available observation wells as we will see in the next section.

5 Simulation in Transient State

The results from the calibration in the steady state will serve as a starting point for the calibration in the transient state. This is why all the elements relating to the hydrogeological functioning and the conceptual model remain valid for the calibration in the transient regime. The assigned data are outlined as follows:

- **Recharge**: This involves the aquifer's replenishment through the inflow of floodwaters from Saguia El Hamra Wadi during specific years (1987, 1991, 2016). Recharge rates were determined annually based on measured data and specific parameters to the model.
- **Pumping Rates for Drinking and Industrial Water Supply**: Pumping rates allocated to pumping points have been assigned to corresponding grid cells.
- **Storage Coefficient**: Given the predominantly sandy sandstone or mildly fractured limestone composition of the aquifer formations, the storage coefficient values are based on few measurements ranging between 10 and 15%.
- **Periods and Calculation Steps**: The starting piezometric condition corresponds to the year 1976, which corresponds to the piezometry achieved at the steady-state and continues until the year 2019. The simulation time is over a period of 44 years, divided into 44 periods of 365 days each (equivalent to one year).

The Foum El Oued aquifer has been under monthly water level monitoring since 1976. The calibration process for the transient model of the Foum El Oued aquifer over a period of 44 years, from 1976 to 2019, with an annual time step (model constraint period = 1 year).

To assess the transient state setup, a set of ten control observation wells were available: 2/120, 5/120, 277/120, 707/120, 734/120, 735/120, 737/120, 739/120, 815/120, and 826/120.

For this study, a uniform spatial distribution of the storage coefficient was adopted. In fact, spatializing this parameter through calibration would have required multiple piezometric records from various points within the aquifer. The selected storage coefficient value is S = 0.1, obtained through an iterative process.

The calculated piezometric data for several representative piezometers are depicted in Fig. 4. The results show a close agreement between calculated and observed heads. Hence, the model effectively reproduces the observed fluctuations' magnitudes and the transient model calibration is generally considered satisfactory. This model enables the reconstruction of the overall aquifer water balance in 2016. Figure 5 illustrates the various components contributing to the calculated balance. However, this transient model is subject to certain limitations that include:

Approximate Distribution of Model Parameters (K, S): The model assumes a uniform distribution with respect to storage coefficient (S). Improvements to this distribution are necessary as new data becomes available.

Simplification of Downstream Boundary Conditions: The specified head condition (H = 0 m) imposed at the sea downstream limit simplifies the model representation at the coast. In fact, the situation is more complex due to the difference in densities of freshwater and seawater. This discrepancy leads to inflows at the boundary instead of a constant outflow as indicated by the zero head condition. This simplification doesn't fully reflect the actual phenomena occurring at this boundary.

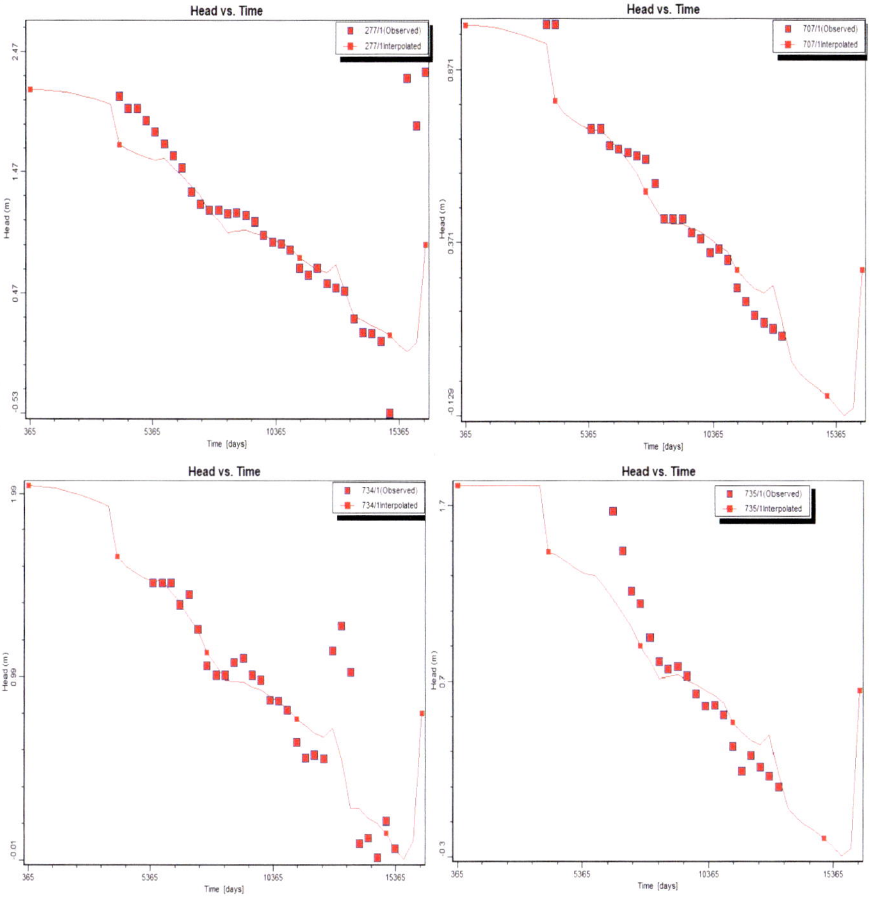

Fig. 4 Transient state model calibration results

6 Seawater Intrusion Modeling

The SEAWAT code intrusion was selected to deal with seawater, it is developed based on the hydro-dispersive transition zone approach. The advantage of using this simulation code in coastal aquifers lies in its integration of two standard modules, MODFLOW and MT3D, adopted by the Moroccan hydraulic administration since the 1990s to deal with saltwater intrusion in coastal aquifers. Furthermore, the data input for this code closely adopts the standard formats of MODFLOW and MT3D codes; and SEAWAT readily accepts existing files and data from previous simulations conducted with MODFLOW. These can be imported and generated with minimal modifications, with the use of pre and post-processing tools like Visual Modflow. Consequently, this code proves to be well-suited for the hydrogeological context and is among robust solutions for addressing seawater intrusion phenomena

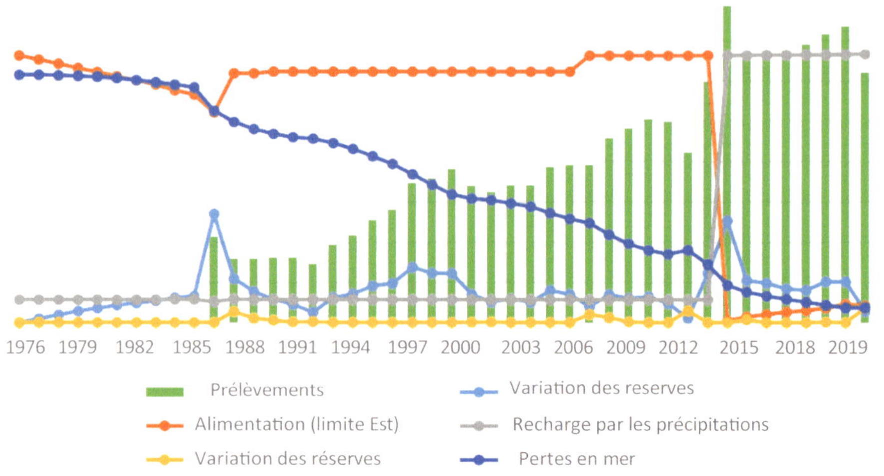

Fig. 5 Water balance showing the main components (1976–2019)

in Moroccan coastal aquifers (Chibout 2019; Chtioui 1986; Draoui 2014; Hamidi 2019; El Mokhtar 2011, 2018; Eyazidi 2004; Fadili 2016; GEO Atlas; Hilali 2002; Kamani 2016; Lakfifi 2002; Zgourdah 2015). Thus, for the preparation of input files and the analysis of model results, we employ Visual Modflow, a software developed by the US Geological Survey.

The variable-density model of the Foum El Oued aquifer retains the overall geometry of the model developed in the transient regime, this means that it includes the following aspects: (i) Land surface topography; (ii) Bottom level; (iii) Lateral boundaries. The design of this model with SEAWAT code (Fig. 6) involves the following specifications: (i) Number of layers = 20; (ii) Layer thickness varies between 6 and 8 m; (iii) The upper boundary of the first layer coincides with the soil surface. Moreover, this simulation is conducted with the following boundary and initial conditions:

- Boundary conditions are equivalent to those used in the transient regime;
- The initial state of the variable-density model corresponds to the steady state;
- Simulation covers a period of 44 years (1976–2019);
- A constant concentration of 35 g/l and a density of 1025 kg/m^3 are assigned to all cells along the specified dimension line within layers 17, 18, 19 and 20.

Within the available Foum El Oued aquifer database, no tracer tests have been conducted in the area. Consequently, the dispersion coefficients were estimate based on references from specialized literature to derive appropriate values, relying on field measurements from lithologies analogous to the aquifer's characteristics. The chosen

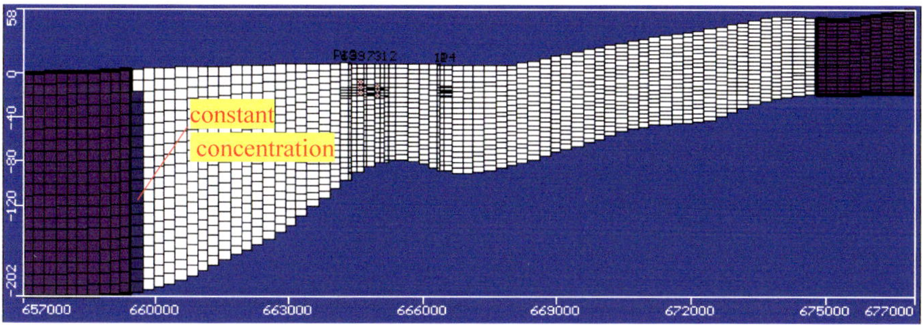

Fig. 6 Cross section displaying grid concentration

value for longitudinal dispersvity stands at 1 m. The ratios of longitudinal dispersvity-transverse dispersvity, as well as longitudinal dispersvity-vertical dispersvity, are set at 0.1 and 0.01 respectively.

At the initial state, corresponding to the initial time step of the variable-density model in 1976, the seawater plume remains within the limits of the western boundary of the Atlantic Ocean. The results show that throughout this period, there are no extraction activities at the capture site. Subsequently, in 1986, the salt plume extends beyond the western boundary of the Atlantic Ocean, aligning with the point when pumping at the capture site was initiated, located 5 km away from the coastline. In the year 2019, simulation results indicate that the current state witnesses the advancement of the saline plume by a distance of 4.5 km, as illustrated in Fig. 7. Furthermore, the iso-concentration line of 10 g/l spans a distance of 4.1 km, while the line representing the highest concentration (35 g/l) reaches a distance of 4 km.

Predictions of seawater intrusion under climate change scenarios

The developed model can serve as a tool for predicting the aquifer's response to scenarios proposed by decision-makers to aid in the planning process and facilitate informed decisions regarding aquifer management. Within this framework, 2 projected climate scenarios (RCP4.5 and RCP8.5) coupled to the GW exploitation were assessed to gauge their potential quantitative and qualitative impacts on the aquifer.

The methodology includes three main steps:

1. *Data Collection for Projected Climate Change (2020–2050)*: Initial step involves gathering projected climate change data encompassing precipitation and temperature variables for the Laayoune area between 2020 and 2050.
2. *Groundwater Recharge Estimation (2020–2050)*: these acquired data are used to estimate groundwater recharge within the period of 2020 to 2050.
3. *Hydro-Dispersive Model Incorporating Sea Level Rise*: Lastly, annual recharge results are integrated into the hydro-dispersive model, accounting for the influence of rising sea levels for the period 2020–2050.

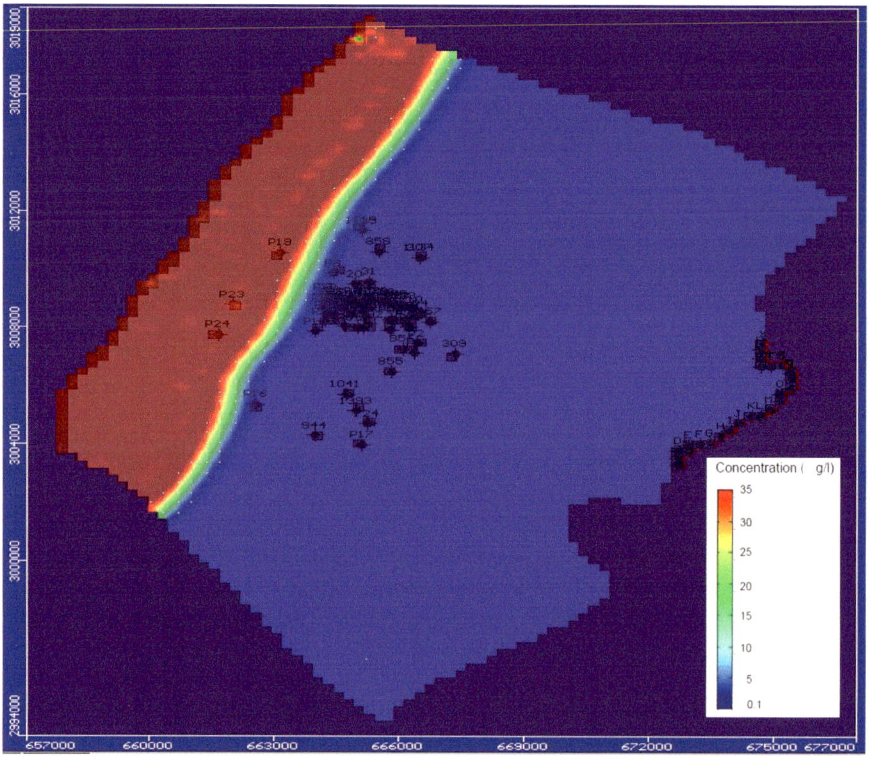

(a)

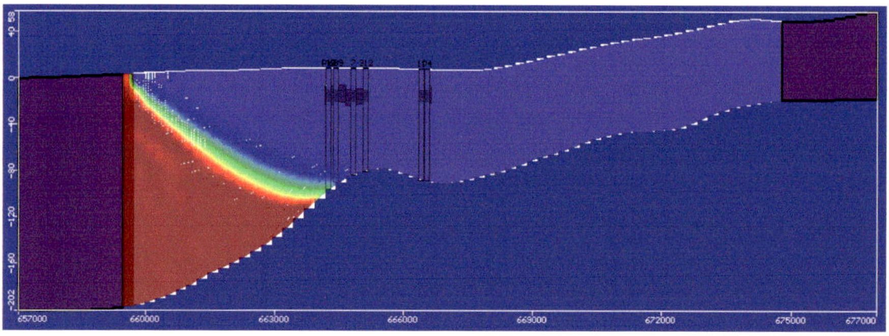

(b)

Fig. 7 Calculated salinity in 2019 **a** salinity distribution in Layer 15 **b** cross-section salinity profile (selected along line 40 of the model grid)

According to the Intergovernmental Panel on Climate Change (IPCC)'s 4th Assessment Report (2007), a projected sea level rise of around 60 cm by 2100 was indicated. Applying a linear trend to our study area, we deduce that sea level will rise from approximately 8.4 cm in 2020 to about 27.8 cm by 2050. This sea level rise is integrated into the seawater intrusion model of the Foum El Oued aquifer.

To simulate future changes in drawdown and salinity concentrations from 2020 to 2050, pumping rates are considered to change based on the management scheme of the Hydraulic Agency (ABH SHOD) in order to satisfy the growing demand.

Simulation results: By 2030, the simulations predict that the salt front will have advanced by around 5.8 km, with the iso-concentration line of 10 g/l extending to 5.5 km. In parallel, the line representing the highest concentration (35 g/l) will have reached approximately 5.1 km.

By 2050, the projections indicate that the salt front will have moved forward by roughly 6.5 km, as depicted in Fig. 8. The iso-concentration line of 10 g/l is predicted to extend to 6.2 km, while the highest concentration line (35 g/l) is anticipated to reach a distance of around 5.8 km.

Salinity evolution and Seawater Intrusion volume: Fig. 9 shows the evolution of calculated salinity. The simulation results suggest a significant salinity increase in the designated area, eventually leading to coastal contamination and abandonment of GW exploitation; because this results in the heightened salt content for water irrigation and water supply from pumping wells (Fig. 10).

Figure 11 illustrates the evolution of seawater intrusion volume from 2020 to 2050 under the RCP8.5 scenario. This depiction reveals a volume increase trend over time, except for some few humid years that would induce natural recharge.

Model validation and Data Limitations: In the absence of continuous withdrawal record data that could enhance the model results, the simulation-derived concentration results are supposed acceptable, particularly in relation to salt front initiation. Salinity-specific measurements from control observation wells for seawater intrusion are scarce, though piezometric measurements, coupled with limited salinity measurements, are available. Hence, comparing model results with salinity measurements lacks significance due to data scarcity. Then, uncertainties still persist concerning pumping and corresponding salinity measurements.

7 Conclusion

The hydrodynamic model of the Foum El Oued aquifer is a three-dimensional (3D) representation that incorporates the complex multilayer geological structure of the system. The model's development was based on the integration of available data that could contribute to the conceptualization of the aquifer system. Due to the lack of certain essential data, some simplifications were necessary to characterize the modeled system. Presently, the mathematical model demonstrates a reasonable level of representativeness with respect to the actual aquifer system.

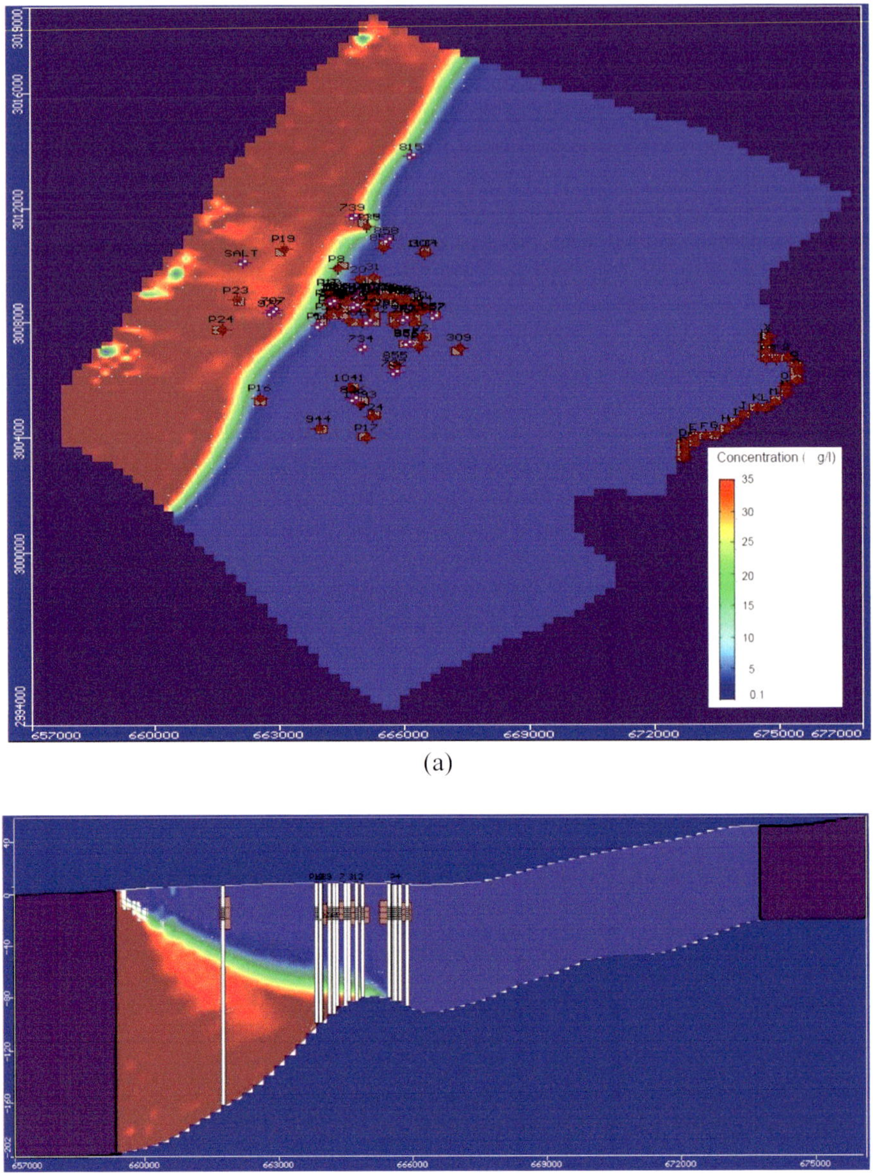

(a)

(b)

Fig. 8 Calculated salinity in 2050 (RCP8.5 scenario) **a** salinity distribution in layer 13 **b** cross-sectional salinity profile (selected along line 42 of the model)

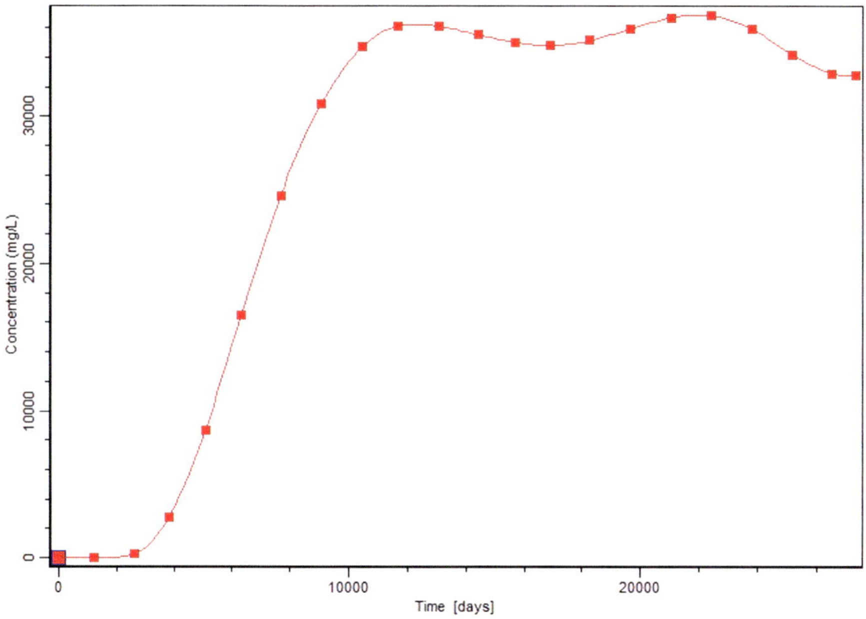

Fig. 9 Evolution of calculated salinity under the RCP8.5 scenario

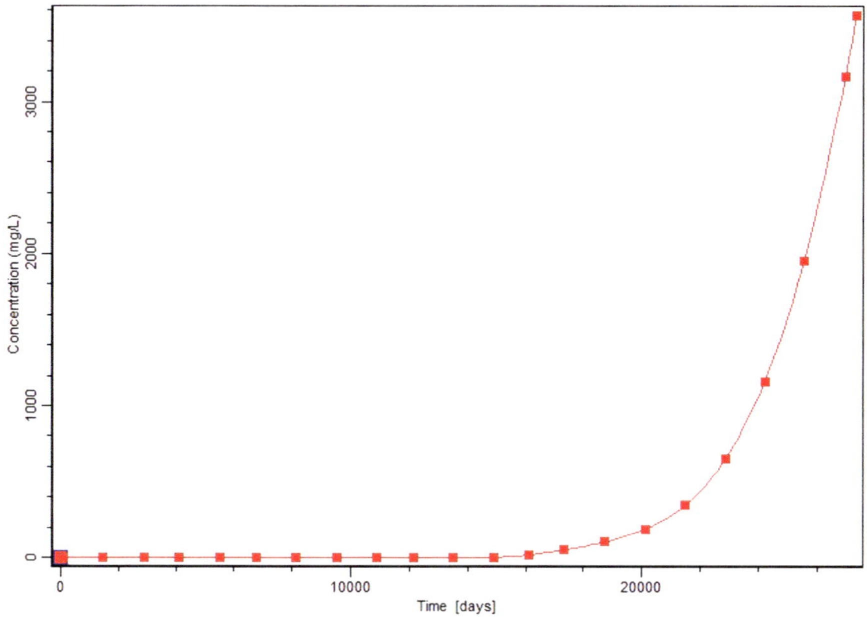

Fig. 10 Evolution of calculated salinity under the RCP8.5 scenario at ONEP well fields

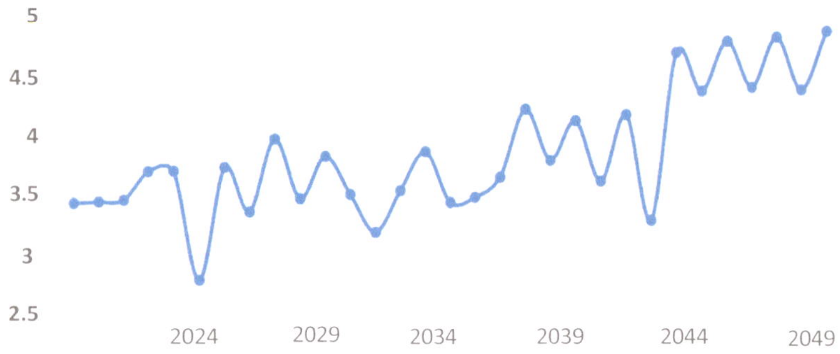

Fig. 11 Seawater intrusion volume in Mm^3 for the RCP 8.5 scenario

However, the model experienced various sensitivity tests, revealing that there exists a notable degree of uncertainty associated with the model's outcomes. As a result, the model should be regarded as a tool that possesses potential for significant refinement.

To achieve a more accurate configuration with the known geological structure of the aquifer, enhancing the 3D model would require differentiation of parameters (such as heads, hydraulic conductivity, storage, etc.) for each individual layer constituting the Foum El Oued aquifer system.

Concluding this study and with the objective of enhancing comprehension of the aquifer system's functioning and its interaction with its surroundings, the following actions are recommended:

- Conduct exploration drilling to determine the northern and southern boundaries of the Foum El Oued aquifer, coupled with accurate leveling and pumping tests.
- Perform additional pumping tests to deduce storage coefficients and acquire comprehensive K-values for the Foum El Oued aquifer.
- Accurately define the extensions of the distinct geological formations composing the aquifer system.
- Undertake well-distributed piezometric campaigns spanning the expanse of the water table, incorporating accurate leveling of observation wells.
- Implement a tracer test to estimate the dispersivity coefficient, thus enhancing understanding of the aquifer's behavior.

By engaging in these proposed activities, a more robust understanding of the aquifer system and its interaction with its environment can be achieved.

References

ABH SHOD (1993) Development of water from the Saguia El Hamra dam

ABH SHOD (2003) Geophysical study by electrical soundings, nuclear magnetic resonance soundings, and seismic surveys in the Laâyoune-Dakhla sedimentary basin. Project No. 03

ABH SHOD (2005) Hydrogeological study of the aquifer complex of the Sahara basin. Project No. 07

ABH SHOD (2009) Project to demarcate the perimeters of the protection and banning of the Foum El Oued water level

ABH SHOD (2010) Inventory of groundwater withdrawals in the Sakia El Hamra and Oued Eddahab basin. Project No. 02

ABH SHOD (2013a) Study of the artificial recharge of the Foum El Oued aquifer, Province of Laâyoune. Project No. 12

ABH SHOD (2013b) Hydrological, piezometric, and spring flow measurement campaigns in the Sakia El Hamra and Oued Eddahab basins, 2011–2013

ABH SHOD (2013c) Salinometry in the areas of Foum El Oued, Laâyoune, 2000–2013

Chibout M (2019) Geophysical contribution to the determination of the limit between Laayoune and Foum El Oued aquifers, origin of supply sources of Oued Sakia El Hamra, Laayoune Province, Morocco

Chtioui M (1986) Hydrogeological study of the Foum El Oued aquifer, assessment and management of water resources. End-of-study project, Mohammadia School of Engineering, Mohammed V University in Rabat

Draoui H (2014) Hydrogeological modeling of the Foum El Oued aquifer (Laâyoune region, Morocco). Final project, Mohammadia School of Engineering, Mohammed V University in Rabat

El Hamidi MJ (2019) Hydrogeological geodatabase, mathematical modeling, and forecasting of marine intrusion under climatic changes in the Rmel aquifer (Morocco). PhD thesis, Mohammadia School of Engineering, Mohammed V University in Rabat

El Mokhtar M (2011) Modeling of the Foum El Oued aquifer using Groundwater Vistas software. End-of-study dissertation, Faculty of Sciences, Ibn Tofaïl University of Kénitra

El Mokhtar M (2018) Assessment of seawater intrusion in the Foum El Oued aquifer, Laâyoune Province, Morocco

Eyazidi R (2004) Modeling and hydrogeology of the Chtouk sub-valley area. End-of-study project, Mohammadia School of Engineering, Mohammed V University in Rabat

Fadili A (2016) Hydrogeological and geophysical study of the extension of seawater intrusion in the Oualidia Sahel (Morocco): statistical analysis, hydrochemistry, and electrical prospecting

GEO ATLAS, Modeling study of the Foum El Oued aquifer (mission: development of a mathematical model for the management of water resources)

Hilali M (2002) Hydrogeology and modeling of seawater intrusion in the coastal aquifers of Martil and Sahel-Morocco. PhD thesis, Mohammadia School of Engineering, Mohammed V University in Rabat

Kamani Y (2016) Hydrogeological modeling of the Foum El Oued aquifer. End-of-study project, Mohammadia School of Engineering, Mohammed V University in Rabat

Lakfifi L (2002) Methodological approach for modeling marine intrusion in coastal aquifers, application to the case of the coastal Chaouia aquifer. End-of-study project, Mohammadia School of Engineering, Mohammed V University in Rabat

Zgourdah Z (2015) Contribution to the study of the salinization of the coastal aquifer of Foum El Oued (Laayoune) in Morocco

Numerical Modeling of the Future Climate Change Impacts on the Ghis-Nekkor Aquifer Under RCP4.5 and RCP8.5 (Al Hoceima, North of Morocco)

Abdelkader Larabi, Hanane El Asri, Mohamed Faouzi, and Mohamed Jalal El Hamidi

Abstract The study area covers the Ghis-Nekkor plain on the Mediterranean coast located in the north of Morocco. This groundwater aquifer is a typical example of a coastal aquifer whose water resources are exposed to severe overexploitation combined with increasingly low water supplies. The Ghis-Nekkor aquifer is an area highly vulnerable to seawater intrusion (SWI), firstly, because of less natural recharge, mean sea level rise due to climate change combined to groundwater overexploitation. The Regional Climate Models (RCMs) indicate a decrease in precipitation (18%) and an increase of average temperature (0.5 °C) in the Ghis-Nekkor area by 2050 under the RCP4.5 scenarios. These changes would have a direct impact on groundwater recharge of the aquifer. The results of the predicted groundwater recharge show a reduction of the mean annual recharge of 14% for (2021–2050) following the RCP4.5 scenario. Moreover, predicted values of future groundwater recharge under the RCP8.5 scenario would decrease more and would reach 30% less for the period (2055–2084). In view of the limited fresh water resources and degradation of water quality by sea water intrusion, effective management of ground water resources in this aquifer is necessary and can be made by developing a groundwater flow model taking into account seawater intrusion. Indeed, the outputs of RCMs for precipitation and SLR were used to assess the impact of pumping and climate change on the Ghis-Nekkor aquifer under the RCP4.5 and RCP 8.5 scenarios (2020–2040). The results show that the N-W sector of the coastal area would be more vulnerable to SWI. By 2040, the length of inland encroachment would reach 1 km, with a significant salinity (25 g/l) and a significant drop in hydraulic head (−15 m) following the RCP4.5 scenario. The extreme case of groundwater recharge and SLR under the RCP8.5 scenario will cause more severe impacts. By 2040, the length of inland encroachment would be around 1.23 km in the N-W part, at a distance of 0.55 km

A. Larabi · H. El Asri · M. Faouzi · M. J. El Hamidi (✉)
Regional Water Center of Maghreb, LAMERN, EMI, Mohamed V University in Rabat, Rabat, Morocco
e-mail: elhamidi@emi.ac.ma

N. Rebai et al. (eds.), *GIS, Applied Computing and Data Science for Water Management*, Lecture Notes in Geoinformation and Cartography,
https://doi.org/10.1007/978-3-031-63038-5_7

from the nearest ONEE well field. The salinity would reach 33 g/l and the SWI volume would increase by 48% (1.78Mm^3/year) compared to the results under the RCP4.5 scenario.

Keywords Ghis-Nekkor · Climate change · Regional climate models · Seawater intrusion · Salinity · Groundwater recharge · Hydrogeology

1 Introduction

Semi-arid regions, such as the Mediterranean basin, are experiencing a growing water deficit due to a combination of increasing demands (resulting from rapid population growth and agriculture), excessive resources exploitation, and the reduction of groundwater resources caused by climate change. The Mediterranean region is particularly vulnerable to climate change, as highlighted in the recent report of the Intergovernmental Panel on Climate Change (IPCC 2013), mainly due to the rapid changes already observed (Giorgi 2006). Temperature increases could potentially reach 3.8 °C by 2100 under the RCP 8.5 scenario, exceeding the global average temperature predictions. Additionally, there may be a reduction in precipitation by 10 to 30%, varying by region (IPCC 2013). These developments will further strain already heavily demanded water resources.

Morocco is one of the semi-arid countries facing water scarcity and prolonged drought episodes (Naciri 1990; Bouaicha and Benabdelfadel 2010). Coastal aquifers in Morocco represent critical groundwater resources for both human activities and the environment. The salinization of these aquifers can be intensified by excessive pumping and sea level rise (SLR) associated with climate change. Positioned geographically near the Mediterranean front to the north, the Ghis-Nekkor area is particularly susceptible to the adverse effects of global warming. Therefore, conducting an assessment of the impact of future climate change on this region is of paramount importance, especially for optimal planning and management of water resources by the regional decision makers.

2 Geology and Extension of the Aquifer

The Ghis-Nekkor plain is situated in the northern region of Morocco (see Fig. 1), and it is entirely bordered by impermeable schistose formations, except for the northern part, which is open to the Mediterranean Sea. From a climatic perspective, the Ghis-Nekkor plain falls within a semi-arid zone characterized by high average temperatures, typically around 18 °C, and low annual rainfall, which does not exceed 300 mm. This plain contains a significant groundwater reservoir, recognized as one of the most substantial in the Rif region. The aquifer primarily consists of Plio-Quaternary alluvium age with a varying thickness reaching approximately 240 m. Its

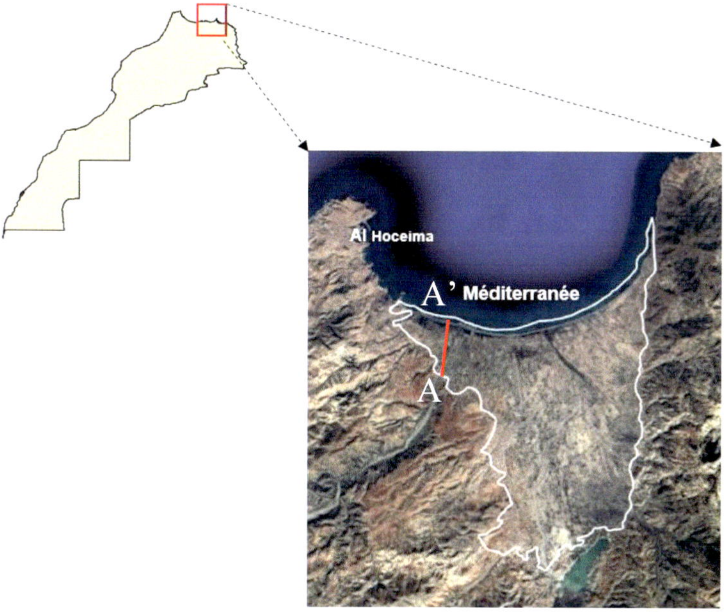

Fig. 1 Map depicting the geographical location of the Ghis-Nekkor plain

schistose bottom gradually slopes from South to North beneath the Plio-Quaternary covering. The aquifer receives its main recharge from the lateral boundaries (about 34%) and from precipitation (around 38%) and subflow from the wadis as they enter the plain (approximately 22%). The Mediterranean Sea serves as the primary outlet for the aquifer, with losses to the sea accounting for approximately 67% of the total groundwater output.

3 Numerical Modeling on the Ghis-Nekkor Aquifer

The process of building the FEFLOW model begins with the generation of a super mesh encompassing the domain (Diersch 2010; DHI-WASY, 2010; DIERSCH H-JG. 2004; DIERSCH HJG. 2006). This generation of the super mesh takes into account the geometric characteristics of the modeled area, including the locations of pumping wells and control piezometers. This approach helps eliminate potential inaccuracies by ensuring accurate positions of reference points on the nodes. The next step involves mesh discretization in two dimensions, with a focus on optimizing the sizes of the triangular elements in the horizontal plane. The resulting two-dimensional mesh is subsequently projected in three dimensions, following the various depth-defined layers, as illustrated in Fig. 2. These defined sublayers enable vertical discretization of the elements, resulting in a mesh consisting of 4 general layers. The model surface elevation represents the highest point within the area, while the model base corresponds to the bottom. In total, the study area contains 8625 triangular elements (2D) leading in total to 6696 nodes.

The three-dimensional model developed in steady state accurately replicated the observed piezometry, the water balance for the reference year (1967), and the spatial hydraulic conductivity distribution. Subsequently, the model was used to simulate seawater intrusion in both a permanent and transient state, with the aim of assessing its quantitative and qualitative impact on the Ghis-Nekkor aquifer. The model was also designed to project and analyse the future trends of seawater intrusion up to 2040. These future trends were simulated while considering the expected SLR under two IPCC scenarios, RCP4.5 and RCP8.5. The predicted simulations were conducted with the following assumptions:

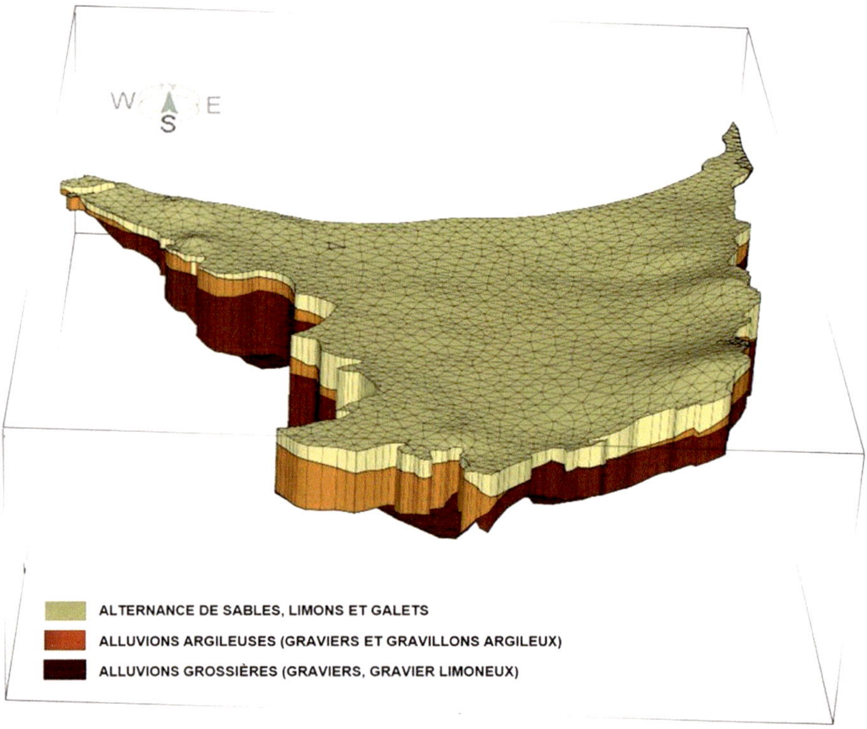

Fig. 2 3D representation of the Ghis-Nekkor aquifer reservoir under FEFLOW

- Agricultural withdrawals remained constant throughout the simulation period.
- The same infiltration coefficient was retained to calculate future recharge from 2021 to 2040 using the MCR HIRHAM5 precipitation projections under both the RCP 4.5 and RCP 8.5 scenarios.
- The infiltration from irrigation surplus from 2020 remained consistent.

The three-dimensional model developed allowed us to examine 4 distinct operational scenarios. The results of the future seawater intrusion simulations under these various forecast scenarios led to the following conclusions.

In general, seawater intrusion was analysed and studied in the Ghis-Nekkor aquifer for four future scenarios.

1. **Scenario 1: Constant pumping + natural recharge (RCP4.5)**

The first scenario, which is the most moderate, aims to simulate the behavior of the groundwater aquifer. In this scenario, groundwater exploitation is modeled using data from the year 2021, with a withdrawal rate of 10Mm^3/year until the year 2040. Aquifer recharge is assessed from precipitation from the MCR (HIRHAM5) under the RCP 4.5 scenario until 2040. The results of the first scenario (see Fig. 3) show a slight increase in the extent of seawater intrusion (0.64 km) compared to the year 2020. The volumes of incoming seawater intrusion would remain relatively close to those recorded in 2020, approximately around 0.35 Mm^3/year by 2040.

2. **Scenario 2: Excessive pumping + natural recharge (RCP4.5)**

The second scenario involves the continuation of aquifer exploitation in accordance with the ONEE pumping projections for 2040 within existing well fields. This is while considering aquifer recharge through rainwater infiltration calculated for the year 2040 from the MCR (HIRHAM5) under the RCP 4.5 scenario. Intensive pumping could lead to an expansion of the area affected by seawater intrusion, especially in the 3rd layer of the aquifer model (refer to Fig. 4). This expansion is estimated to be approximately 0.81 km along section AA', situated to the northwest of the aquifer and intercepting the ONEE well field (Fig. 1). Under this scenario, the results indicate a rising trend in the volumes of seawater intrusion invading the aquifer. By 2040, the

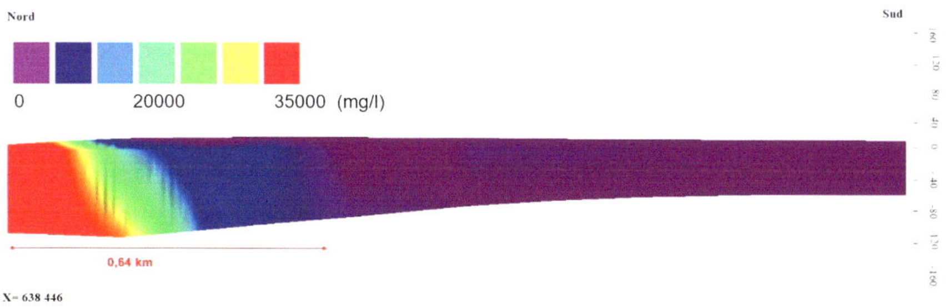

Fig. 3 Salinity and extent of seawater intrusion along cross section AA' for Scenario 1 by 2040

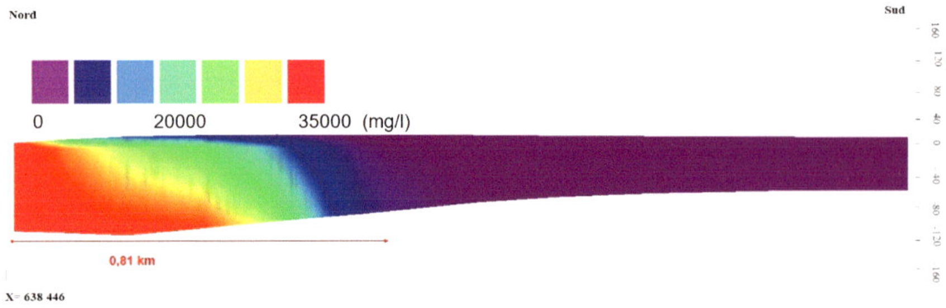

Fig. 4 Salinity and extent of seawater intrusion along cross section AA' for Scenario 2 by 2040

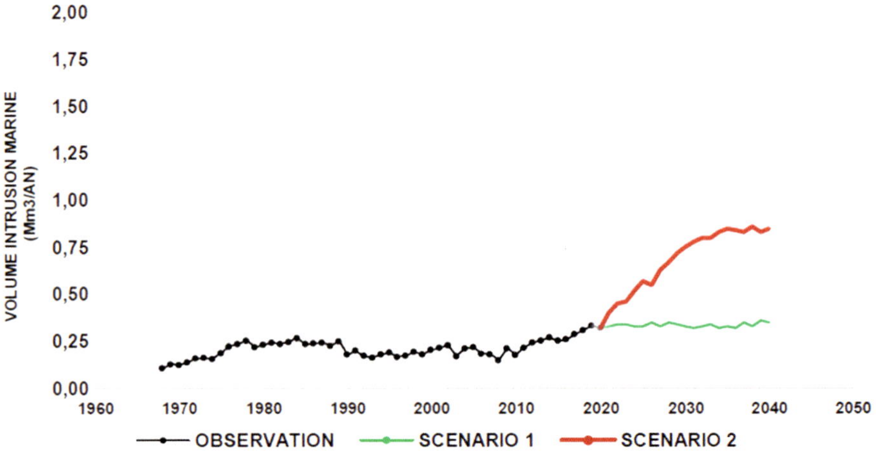

Fig. 5 Evolution of seawater intrusion volumes for the observation period (1967–2020) and the future period (2021–2040) under the two projected scenarios

volume of seawater intrusion is projected to be around 0.85 Mm^3/year (see Fig. 5). Consequently, salinity levels would exceed 25 g/l in the northwestern part of the Ghis-Nekkor aquifer (see Fig. 6).

3. **Scenario 3: Excessive pumping + recharge (RCP4.5) + Sea Level Rise (SLR) RCP4.5**

In this scenario, in addition to the assumptions made for the second scenario, the impact of the SLR was taken into account for 2040 under the RCP4.5 scenario, corresponding to a rise of 0.18 m (Aral and Chang 2017). The simulation results (refer to Fig. 7) indicate that the combined effect of SLR and intensive pumping could have a severe impact on the Ghis-Nekkor aquifer. The northwestern part of the aquifer would be particularly affected, with a substantial advancement in the extent of seawater intrusion observed across all three layers of the model (see Fig. 8). Salinity levels would exceed 30 g/l (as shown in Fig. 9), and the expansion of seawater intrusion

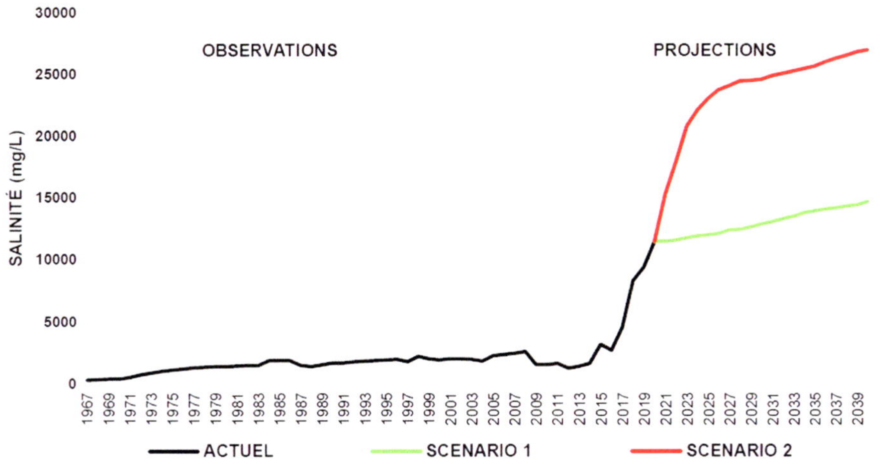

Fig. 6 Evolution of salinity in the third layer of the model at point 1684/5 for the observation period (1967–2020) and the future period (2021–2040) under the two projected scenarios

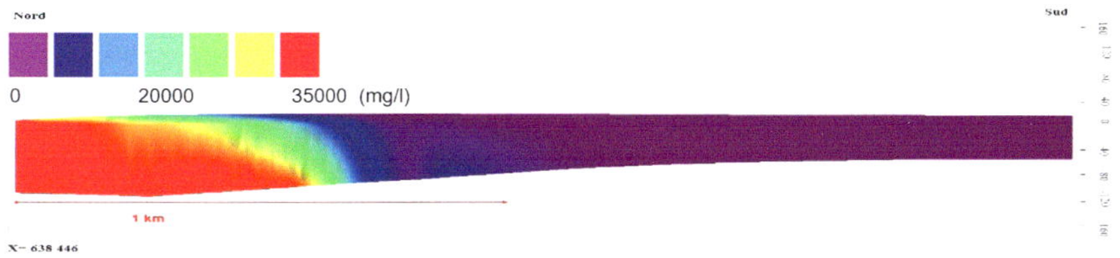

Fig. 7 Salinity and extent of seawater intrusion along cross section AA' for scenario 3 by 2040 (SLR + 0.18m projected for RCP4.5 scenario)

would reach approximately 1 km at the third layer of the model, representing a 23% increase compared to the results of scenario 2. Intensive pumping, in conjunction with a potential SLR, would result in a significant decline in hydraulic head levels, especially in the northwestern section of the aquifer, where drawdowns are projected to reach negative values of around −10 m by 2040 (see Figs. 10 and 11). In this scenario, the volume of seawater intrusion would increase and it is estimated to be around 1.2 Mm^3/year.

4. **Scenario 4: Excessive pumping + recharge (RCP8.5) + SLR (RCP8.5)**

When considering the same ONEE pumping projections for 2040, the simulation of future seawater intrusion was conducted under the IPCC RCP 8.5 scenario. Future recharges were calculated using precipitation data from the MCR (HIRHAM5) under the RCP 8.5 scenario, with a SLR of approximately 0.22 m expected by 2040. The simulation results (see Fig. 12) reveal an expansion in the extent of seawater intrusion, which could reach 1.23 km along section AA'. The anticipated reduction in recharges for 2040, coupled with intensive pumping and the SLR under the extreme RCP 8.5

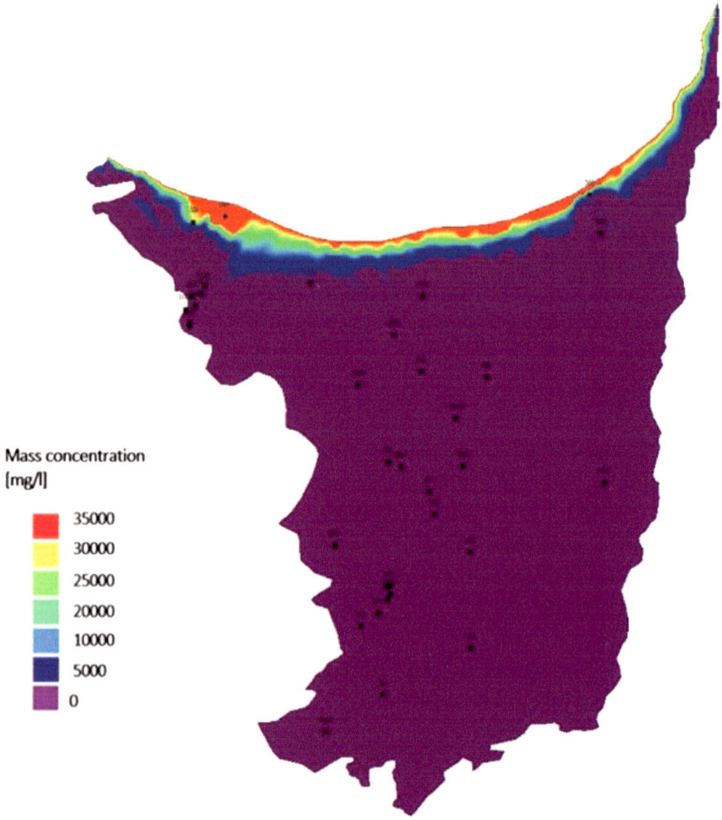

Fig. 8 Map depicting the spatial distribution of simulated salinity for 2040 under scenario 3 in the 2nd and 3rd layers

scenario, would significantly impact both the quantity and quality of fresh water available in the aquifer. The results indicate a substantial increase in the volumes of seawater invading the aquifer, projected to reach 1.78 Mm^3/year, representing a 48% increase compared to the results of the RCP 4.5 scenario, with salinity levels trending towards 34,000 mg/L. Furthermore, drawdowns are expected to reach more negative values of approximately −20 m by 2040 (as depicted in Figs. 13 and 14). In this scenario, the volume of seawater intrusion is estimated to be around 1.2 Mm^3/year. Table 1 provides a summary of the saltwater intrusion volumes invading the aquifer under various scenarios, in addition to the salinity concentrations and the SWI fronts.

An analysis of the summary of the results reveals that, in general, seawater intrusion occurs in the Ghis-Nekkor aquifer across all four scenarios, leading to an expansion of the contaminated front and a reduction in the distance between the nearest ONEE well and the saltwater interface. By 2040, this distance could potentially

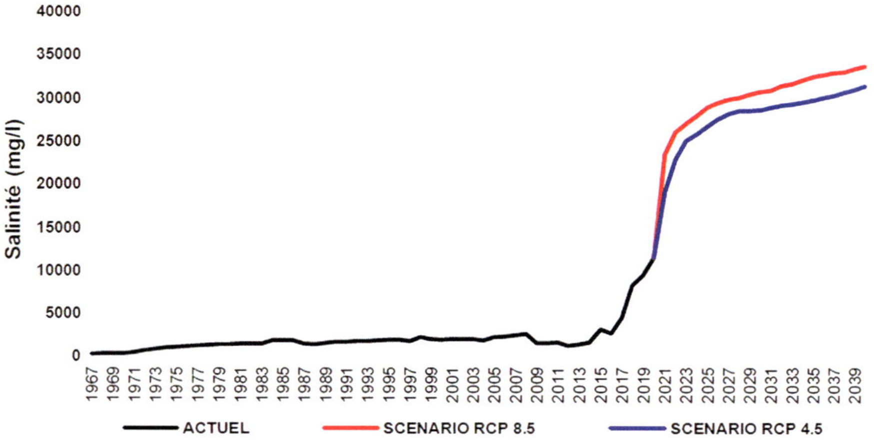

Fig. 9 Future evolution of salinity in the third layer of the model at observation well 1684/5 under the two projected scenarios (scenarios 3 and 4)

decrease to 0.5 km in accordance with the RCP8.5 scenario, with salinity levels exceeding 30 g per liter in the lower model layers. Future water balances were estimated for the least projected rainy years, specifically in 2035 for the RCP4.5 scenario and in 2038 for the RCP8.5 scenario (see Table 2). Over a period of 53 years (1967–2021), groundwater water inflows decreased by 18%. During over half of this period, inflows saw an additional 10% reduction under the RCP4.5 scenario and a 23% decrease under the RCP8.5 scenario. The deficit would accelerate according to both scenarios, but it is more pronounced for the RCP8.5 scenario. Furthermore, under the RCP8.5 scenario, the volume of freshwater inputs is projected to be 10 Mm^3 by 2038, which is 52% less compared to the year 1967. An extreme situation of pumping combined with SLR, along with the arid conditions that the aquifer would experience by 2080 (IA = 9.61mm/°C), would certainly lead to even more adverse effects. Figure 15 illustrates the storage balance along the coastline for transient conditions during the period 2021–2040, while Fig. 16 shows the evolution of seawater intrusion volumes for the observation period (1967–2020) and the projected period (2021–2040) under three scenarios.

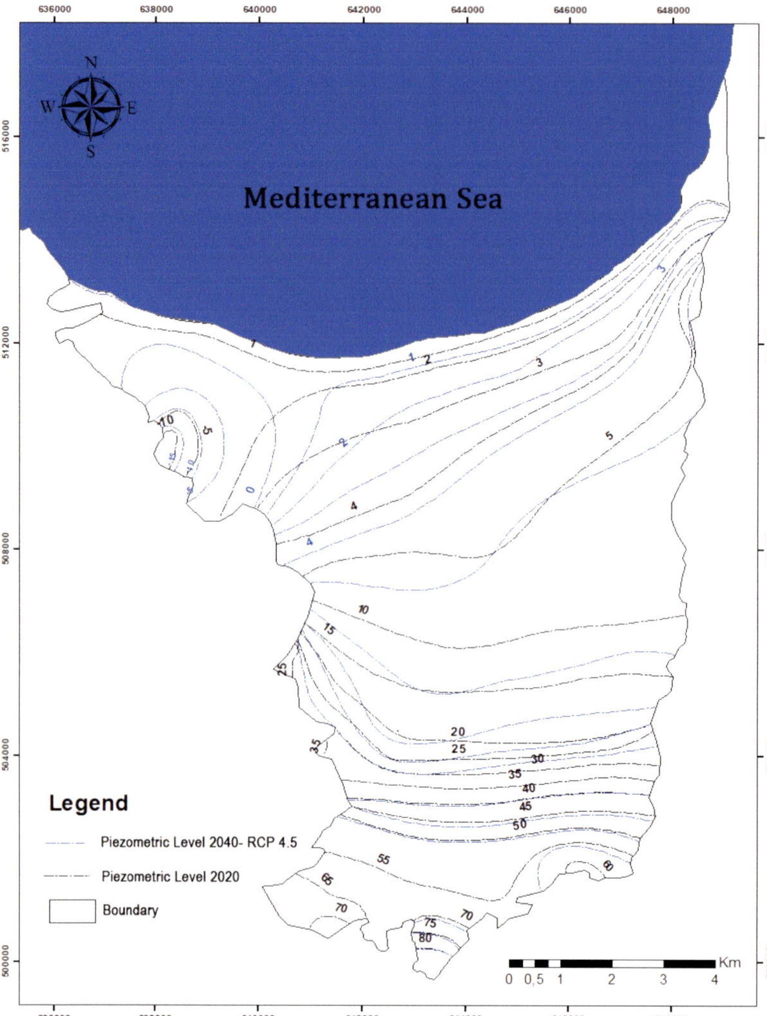

Fig. 10 Map comparing piezometric levels for the year 2020 and the year 2040 for scenario 3

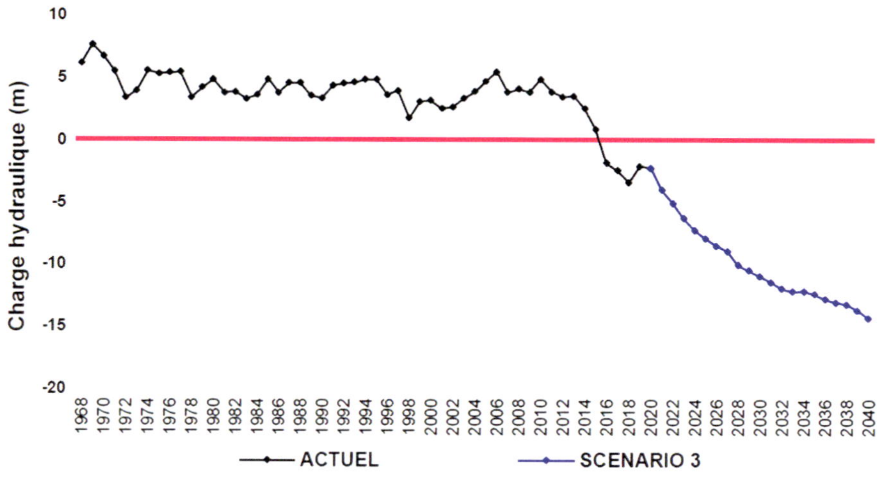

Fig. 11 Future piezometric records at well 385/5 in the ONEE well field

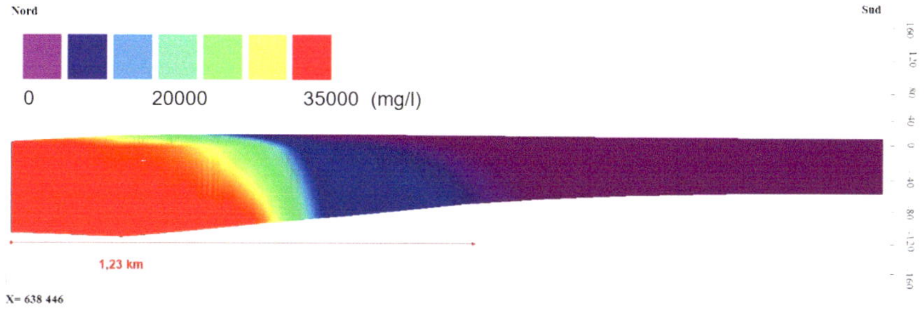

Fig. 12 Salinity and extent of seawater intrusion along cross section AA' for scenario 4 by 2040

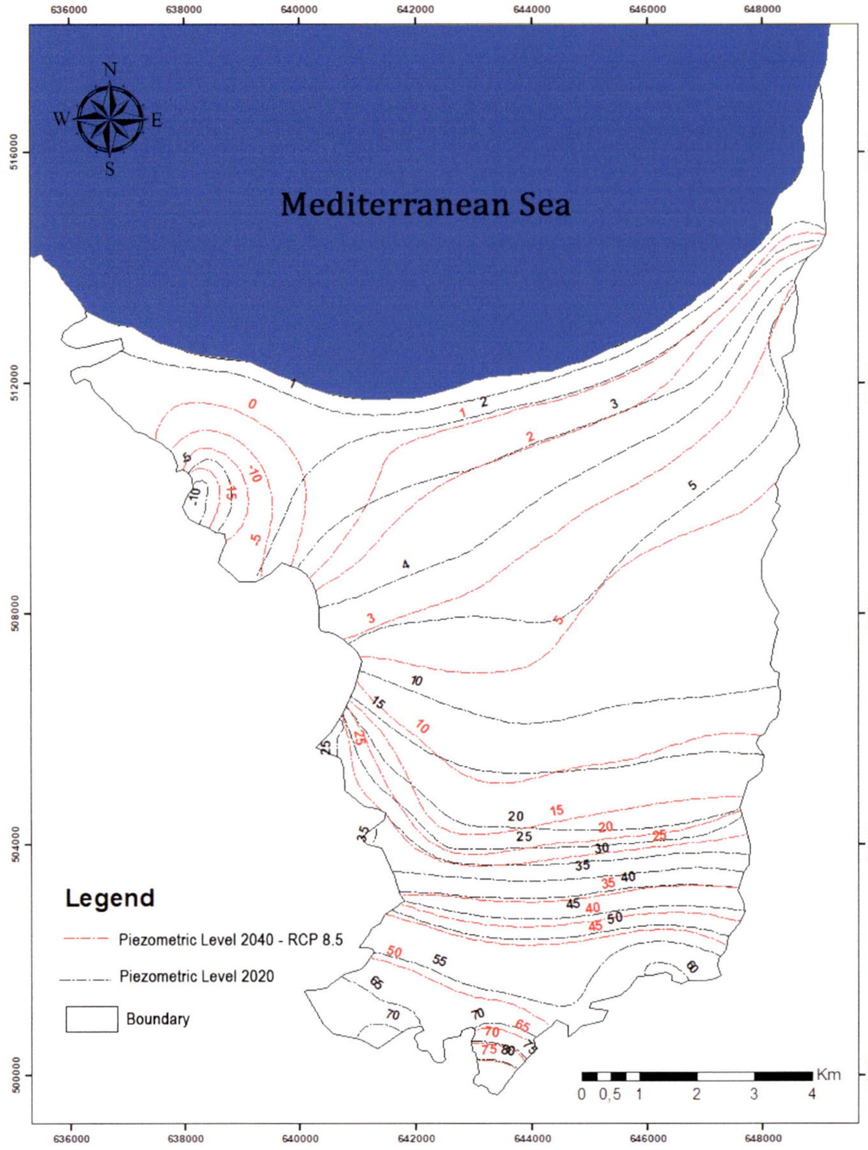

Fig. 13 Map comparing piezometric levels for the years 2020 and 2040 (sc4)

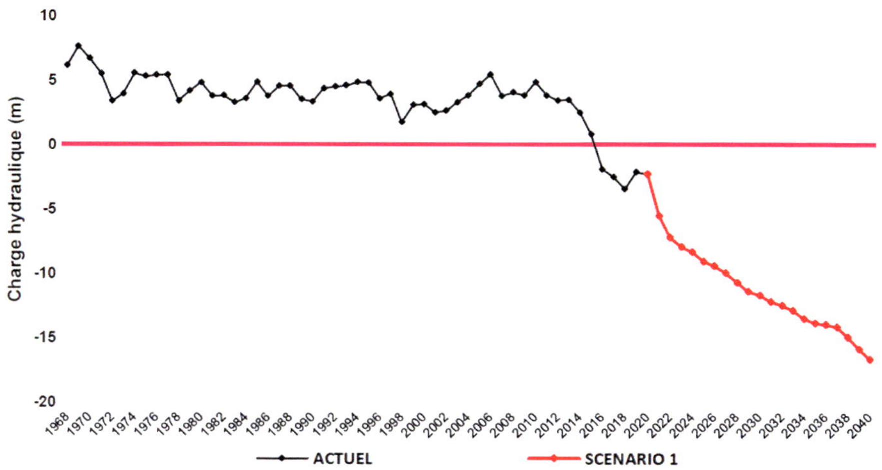

Fig. 14 Future piezometric records at well 385/5 in the ONEE well field (sc4)

Table 1 Summary of projected salinity results and the extension of the seawater intrusion front under the two IPCC scenarios, RCP4.5 and RCP8.5

Scenario	Volume SWI (Mm^3/an)	Extension of the SWI front (km)	Salinity (mg/l) NW (1684/5)	Distance from the nearest ONEP borehole % to the salt bevel (km)
1	0.34	0.64	22,700	1.23
2	0.81	0.81	27,000	1
3	1.2	1	32,080	0.81
4	1.78	1.23	33,578	0.5

Table 2 Evolution of future water balances

Balance components	1967	2021	2035 RCP4.5	2038 RCP8.5
Storage	0	0.5	2.01	3.52
Lateral flow	6.92	4.97	3.45	2.5
Seawater intrusion	**0.11**	**0.34**	**1.2**	**1.78**
Irrigation returns	1.24	0.66	0.66	0.66
Recharge by precipitation	8.23	5.6	3.86	3.25
Subflow contributions	4.5	5.85	4.45	3.73
Total	**20.97**	**17.92**	**15.63**	**15.45**
Destocking	0	0.03	0	0.02
Pumping	2.23	9.8	13.91	14.65
SVM	14.3	3.77	0.43	0.12
Evaporation	4.33	3.95	1.2	0.6
Total	**20.86**	**18.02**	**15.54**	**15.39**

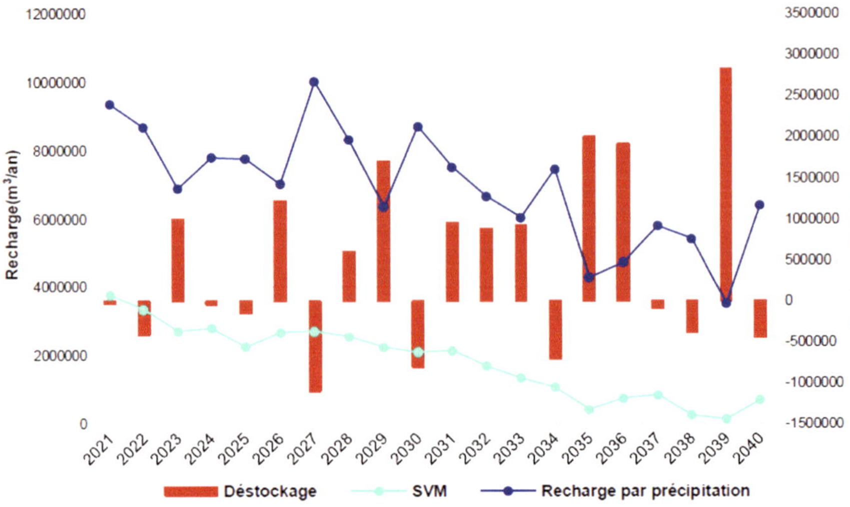

Fig. 15 Water balance components under transient conditions for the period 2021–2040—balance of storage crossing the coast

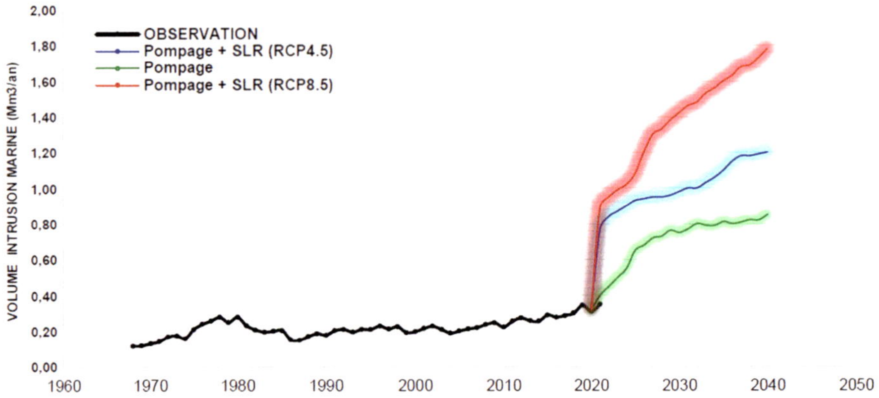

Fig. 16 Evolution of seawater intrusion volumes for the observation period (1967–2020) and the future period (2021–2040) under three scenarios

4 Conclusion

It is evident that intensive pumping is the primary driver of seawater intrusion advancement in the Ghis-Nekkor aquifer. The northwestern part, due to its proximity to the Ghis-Nekkor ONEE well field, is particularly vulnerable to seawater intrusion. The SLR, in line with the two IPCC assessment scenarios (RCP4.5 and RCP8.5), when combined with increased pumping, could further intensify the already concerning situation in the era of climate change.

The results obtained highlight the impact of climate change alongside the deliberate overexploitation of the Ghis-Nekkor aquifer. Seawater intrusion is spreading at an alarming rate, particularly in the case of the severe RCP 8.5 scenario. In an extreme scenario, where pumping, SLR, and the arid conditions, the aquifer may experience by 2080 converge, the adverse effects which are bound to intensify.

Therefore, it is clear that the implementation of a proactive policy with preventive measures and actions is urgently needed to maintain the groundwater aquifer equilibrium, especially in face of climate change. This policy should encompass the following steps:

1. Conduct comprehensive pumping tests covering the entire Ghis-Nekkor area to estimate hydraulic conductivities, storage coefficients, and dispersive coefficients.
2. Explore additional alternative water resources, such as artificial groundwater recharge, to compensate for potential deficits.
3. Consider the establishment of a desalination plant to reinforce the supply of drinking water, using seawater, particularly for the DWS (Drinking Water Supply).
4. Enhance the piezometry and salinity monitoring network and regularly track its evolution, while relocating wells within the well fields.
5. Utilize advanced climate models, especially those focusing on the region, such as the Europe CORDEX models, to assess the impact of SLR with refined mesh resolution in the area.

Implementing this preventive policy will significantly enhance groundwater quality in coastal areas and protect freshwater resources from seawater intrusion. Importantly, such a policy must be part of a broader, global effort. While local management of aquifer exploitation is essential, SLR is a global phenomenon influenced by international policies. Morocco should, therefore, take a national leadership role in addressing overexploitation of aquifers while actively participating in international efforts to reduce greenhouse gas emissions, aligning with recommendations stemming from COPs and global initiatives to combat climate change.

References

Aral MM, Chang B (2017) Spatial variation of sea level rise at Atlantic and Mediterranean Coastline of Europe. Water 9 :522. https://doi.org/10.3390/w9070522

Bouaicha R, Benabdelfadel A (2010) Variabilité et gestion des eaux de surface au Maroc. Sécheresse En Ligne 21(1e):325–326. https://doi.org/10.1684/sec.2010.0276

DHI-WASY (2010)FEFLOW 6: user's manual. DHI-WASY, Berlin

Diersch H-JG (2004) FEFLOW: finite element modeling of flow, mass and heat transport in porous and fractured media. Springer, Heidelberg

Diersch HJG (2006) FEFLOW finite element subsurface flow and transport simulation system-reference manual. User's manual, White Papers-Release 5.3, DHI-WASY, Berlin

Diersch H (2010) FEFLOW user's manual. WASY, Berlin

Giorgi F (2006) Climate change hotspots. Geophys Res Lett 33:L08707. https://doi.org/10.1029/2006GL025734

IPCC (Intergovernmental Panel on Climate Change) (2013) Working group II contribution to the Forth Assessment Report of the Intergovernmental Panel on Climate Change

Naciri M (1990) Calamites naturelles et fatalités historique. Science Et Changements Planétaires Sécheresse 1(1):1–16

Managing Groundwater Withdrawal Using a DSS Based on GIS-MODFLOW Coupling Tool: A Case Study of the Berrechid Aquifer, Morocco

Adil Zerouali, Mohamed Jalal El Hamidi, Abdelkader Larabi, Mohamed Faouzi, and Omar Chafik

Abstract The hydrodynamic modeling of the Berrechid aquifer using the MODFLOW code enables the understanding of the aquifer's hydrodynamic functioning, including the estimation of lateral exchanges between the aquifer and the Settat plateau, Oued Mellah, and the Chaouia coastal aquifer. Calibration of the mathematical model under steady-state conditions using piezometric state at 1980 enhances the spatial distribution of hydraulic conductivities across the entire domain and enables an assessment of the aquifer resevoir under steady-state conditions. While calibration of the model under transient conditions refines the spatial distribution of the aquifer's storage coefficient. Results from this simulation indicate a water deficit, with an average drawdown of 2 m per year. Recent droughts in the area have significantly impacted water supplies for both the population and agriculture, underscoring the need for a tool to manage water withdrawal and intervention directly at user levels to protect groundwater against overexploitation. A management groundwater withdrawal tool has been developed for the Berrechid aquifer, using a Decision Support System (DSS) based on coupling ArcGIS to MODFLOW to calculate drawdown caused by groundwater withdrawals at specific areas.

Keywords Hydrodynamic modeling · Berrechid aquifer · MODFLOW code · Steady-state conditions · Transient conditions · Management water withdrawal tool · ArcGIS · Drawdown

A. Zerouali · M. J. El Hamidi (✉) · A. Larabi · M. Faouzi
Regional Water Center of Maghreb, LAMERN, EMI, Mohamed V University in Rabat, Rabat, Morocco
e-mail: elhamidi@emi.ac.ma

O. Chafik
Bouregreg Hydraulic Basin Agency, ABHBC, Benslimane, Morocco

N. Rebai et al. (eds.), *GIS, Applied Computing and Data Science for Water Management*, Lecture Notes in Geoinformation and Cartography,
https://doi.org/10.1007/978-3-031-63038-5_8

1 Introduction

Based on previous studies gathered from the Agence du Bassin Hydraulique du Bouregeg et de la Chaouia (ABHBC) and earlier research (ABHBC 2004, 2016; Water resources of Morocco 1975; MTPC 1977; Mansouri 1993; DRPE 1997), the Berrechid plain, covering a total area of 1500 km^2, is located to the south of Casablanca (see Fig. 1). It takes the shape of a vast basin, bordered by the Settat plateau to the south and southeast, the Oued Mellah valley to the northeast, primary outcrops to the west and northwest, and the Chaouia aquifer to the north.

The plain's morphology is relatively uniform, with NE-SW oriented contours dominating most of the area. Elevations vary from 350 m at the boundary with the Settat Plateau to 140 m in the northern transition zone towards the Chaouia plain. The topographic slope is generally below 0.2% in the central and western sectors but can reach 0.8% in the southeast along the border with the Settat Plateau.

The agricultural focus of the Berrechid Plain has led to increased groundwater utilization. This situation is exacerbated by recurring droughts since the 1980s. Furthermore, the city of Berrechid hosts a significant industrial zone with robust industrial activities, resulting in growing water demands. Hence, the expansion of groundwater pumping has led to substantial overexploitation of the aquifer system.

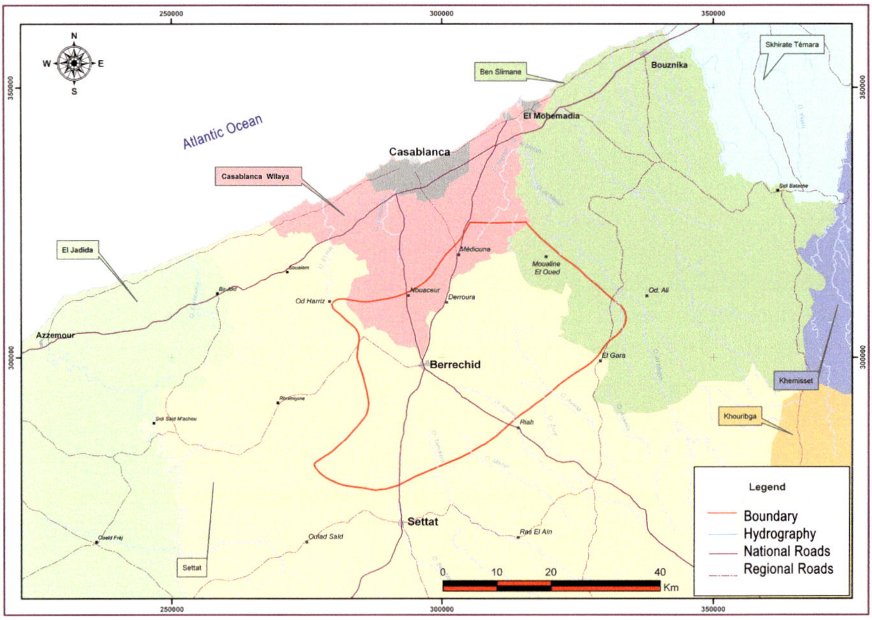

Fig. 1 Location of the Berrechid plain to the south of Casablanca city

2 Hydro-Climatological, Geological and Hydrogeological Context

The climate of the Berrechid Plain is semi-arid, with an average annual rainfall of 395 mm and an average temperature of 28 °C (reaching 45 °C in summer and 10 °C in winter).

The Berrechid Plain is an endorheic area with no functional streams (wadis) draining its hydrographic network. Therefore, the network is poorly developed and mainly consists of some wadis originating from the Settat highland. These streams, lacking outlets to the sea, cross the plain for less than 20 km before diminishing and disappearing in the plain as illustrated in Fig. 1. They exhibit temporary flow, with their partial water quantoties either evaporating or percolating towards the aquifer system. Most important streams are Asseïla, Aiada, Ahmer, Mazer, and Tamdrost, contributing significantly to the natural recharge of the Berrechid aquifer system. Their contributions are particularly important during floods, as demonstrated by rapid rises of the groundwater table observed in several piezometers, located in the area, shortly after such events.

For the geologic setting, tectonics in the study area are relatively straightforward. The most significant tectonic movements occurred during the Hercynian period, resulting in the folding of primary rocks. Subsequent vertical movements, including uplifts and subsidence, formed large blocks, including the coastal Meseta area. Normal faults or regional extension flexures often delineate boundaries between collapsed blocks. These vertical movements influenced sedimentation and erosion within the Meseta's influence area during the Secondary and Quaternary periods. Geological formations encountered in the Berrechid Plain, from bottom to top, include:

- Schists, quartzites, and dolomitic limestones of the Primary period;
- Clays, basalts, conglomerates, and red sandstones of the Triassic period;
- Cenomanian marly limestone;
- Sandy limestone, sand, sandstone, and lumachellic silt of the Plio-Quaternary period.

Groundwater flow occurs within various soil types, which can be:

- Slices and altered network of cracks in the schists and quartzites of the Primary, few levels of Triassic and Infracenomanian.
- Cenomanian marly limestone formations.
- Pliocene sandstone aquifer formations.
- Alternating conglomeratic Quaternary formations.

We distinguish the primary aquifer which is located beneath Quaternary silts, and consists of Pliocene sandstone. Along the edges, and in certain small areas, the sandstone bedrock rises, disrupting the continuity of older horizontal horizons.

The average thickness of the aquifer is approximately 20 m and can exceed 30 m to the right of the hydrographic network of fossil channels from the pre-Pliocene era.

The evolution of the piezometric records indicates a stable water table showing an equilibrium from the 1950s until 1980, followed by a period of groundwater deficit extending from 1981 onwards. This decline in the water table can be attributed to the combined effects of low recharge rainfall and increased groundwater withdrawals, particularly during drought periods.

Groundwater quality in the Berrechid aquifer is generally poor due to high mineralization and nitrate concentrations, mainly from agriculture activities. However, groundwater in this aquifer system exhibits high organic and bacteriological quality.

The intensive groundwater exploitation of the aquifer, coupled with less natural recharge due droughts, has resulted in:

- A decrease in groundwater levels of up to 18 m in some locations, with an average annual decrease of 2 m.
- Degradation of groundwater quality.

The volume of groundwater withdrawals is estimated at 90 Million m^3/year. Prior to 1980, the balance between groundwater abstraction and natural recharge was relatively balanced (Water resources of Morocco 1975). However, in 2003 groundwater balance (total inflow: 33 Mm^3, total outflow: 90 Mm^3) indicates a significant deficit of 57 Mm^3 due to excessive groundwater pumping in the aquifer associated to a rainfall deficit of approximately 140 mm/year.

3 Groundwater Modeling of the Berrechid aquifer

Hydrodynamic modeling of the Berrechid aquifer is conducted using numerical simulation software for groundwater flow, MODFLOW (Macdonald and Harboaugh 1988) with the GMS pre and post processor. The domain discretization consists of 266 columns and 166 rows, delineating 44,156 regular square mesh cells as depicted in Fig. 2 with a side length of 250 m, including 24,401 active cells. The mesh size was chosen considering the density and variability of available data. A 33° rotation of the coordinates of the modeled domain is performed to minimize the number of inactive cells.

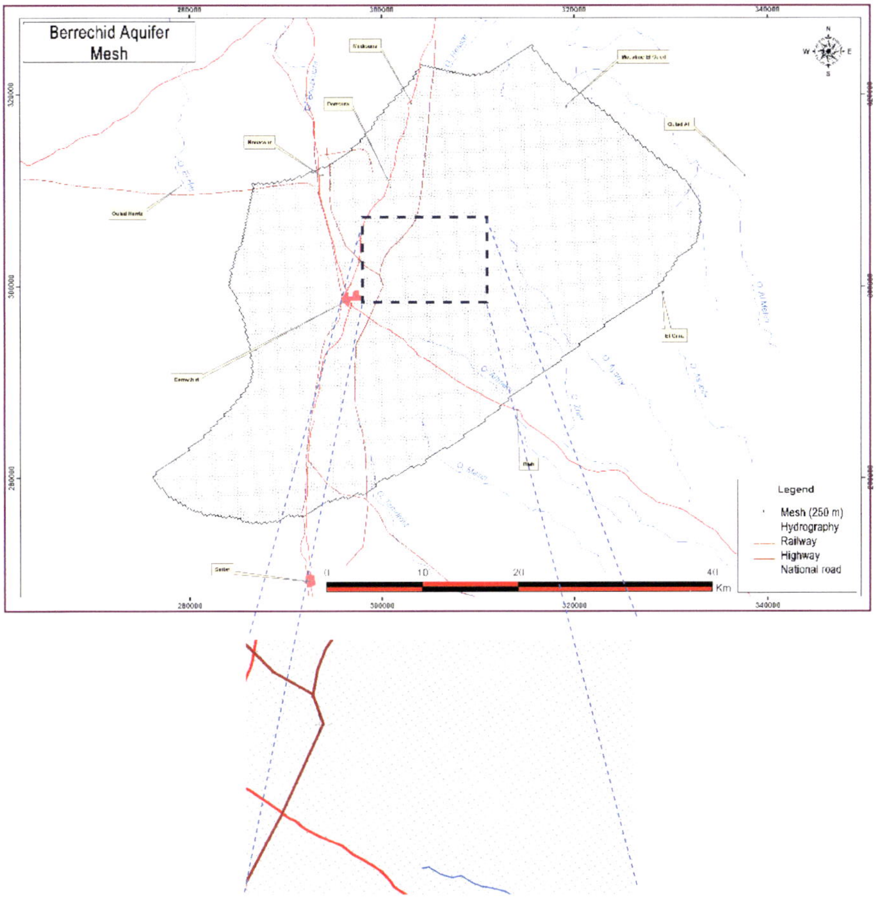

Fig. 2 Mesh of the study area with finite difference cells

The boundary conditions adopted in the model design are as follows:

- specified head of supply along the edge of the aquifer system and the Settat plateau,
- specified head of supply in the interactions between the aquifer and along the wadis from the phosphates plateau (Tamdrost El Ahmeur, Mazère, and Aiada Asseïla),
- imposed drainage potential at the edge of the Berrechid aquifer and the coastal Chaouia aquifer,
- specified head draining areas on the eastern boundary of the aquifer system with the valley of the Oued Mellah where slight drainage is possible,
- a limit with no flow at the Primary schistous soils in the West,

- and a limit of no flow in some areas in the east where clayey soils along the valley of the Oued Mellah are highly collected.

Calculating the infiltrated water quantity presents a challenge due to the need for accurate consideration of the recharge area. In the plain, silt covers an area of 280 km^2, which means that the groundwater aquifer is confined. Therefore, the contribution from rainfall will be calculated in the sector where the groundwater aquifer is unconfined, which is approximately estimated to 1220 km^2. The results indicate an average natural recharge of about 14 Mm^3. This recharge varies depending on rainfall and water availability in different years, ranging from one hundred million m^3 in 1996/97 to almost zero during very dry years such as 1981/82 and 1994/95.

Groundwater withdrawals in the Berrechid aquifer play an essential role in irrigation, drinking and industrial water supply. Groundwater pumping for irrigation and the evolving needs of agriculture in the area have led to continued growth in groundwater pumping. This increasing trend is particularly pronounced during times of droughts and dry periods. For instance, the groundwater volume withdrawn in 1980 was 36 million m^3 (Water resources of Morocco 1975).

4 Steady-State Calibration

The piezometric reference state selected for calibrating the model under steady-state conditions is the piezometric map of 1980, corresponding to the onset of groundwater level drawdown. The map (Fig. 3) depicts the hydraulic conductivity distribution obtained at the end of calibration, revealing distinct zones:

- an area characterized by very low permeabilities, with values below 10^{-4} m/s, mainly situated south of the area. These sectors represent regions where groundwater flow is less transmissive due to thin fringes of sandstone. These zones correlate with the strongest hydraulic gradients observed, highlighting the relationship between the spatial distribution of permeability and the overall piezometer pattern.
- locations with good permeability, with values exceeding 10^{-3} m/s, primarily located in the west and southwest of the area.
- sectors with exceptional permeability, exceeding 10^{-2} m/s, attributed to special drilling locations characterized by coarse fillings at the sites of certain drainage channels from the Antepliocene period.

The remaining area exhibits intermediate permeability, ranging between 10^{-4} and 10^{-3} m/s. Figure 4a overlays the calculated piezometric levels from the model with those observed in 1980, demonstrating satisfactory agreement in terms of calibration. Figure 4b also shows a good agreement between the observed and calculated drawdowns in the control observation wells and demonstrated by a very good correlation coefficient of 99%.

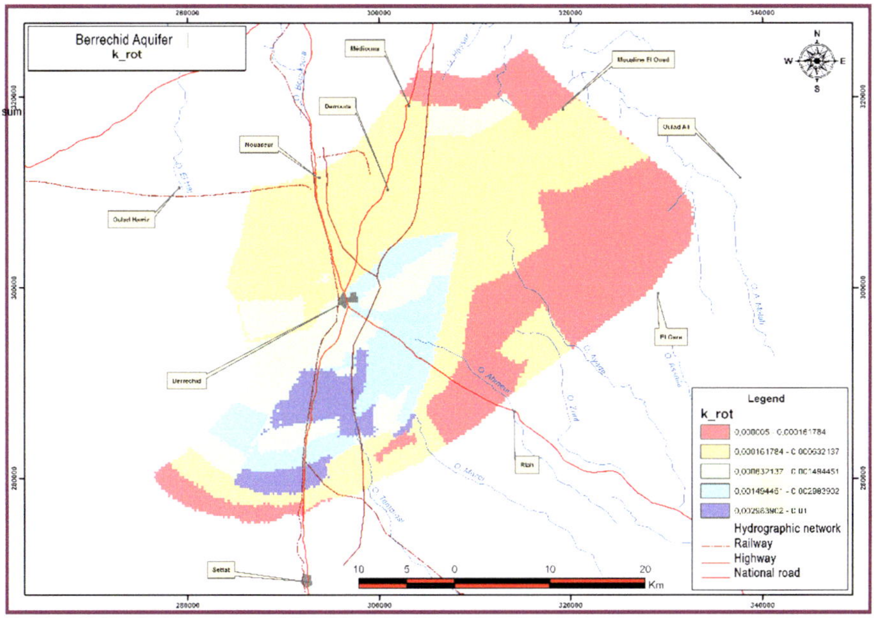

Fig. 3 Calibrated K-values map

This calibration of the model under steady-state conditions has enabled the reconstruction of the overall groundwater balance and assessment of its various components (refer to Table 1). The overall balance obtained from the model mainly agrees with that assessed from the hydrogeological synthesis study conducted in 1980 (Water resources of Morocco 1975).

5 Simulation in Transient Conditions

This section, dedicated to the model under transient conditions, represents a continuation of the steady-state step and considered as the initial conditions for the transient model. The primary objective of this stage of hydrodynamic modeling is the calibration of the storage coefficient and the homogenization of transient regime data within the aquifer system. The selected simulation period spans 31 years, starting from 1980, and is divided into equal-duration intervals of one year.

The distribution of effective porosity in the studied area was derived from available data, with six values ranging between 0.5 and 3.5% (Water resources of Morocco 1975). This distribution is being adjusted during the calibration process to fit the piezometric records with the calculated drawdowns in the control observation wells.

Various variable terms related to inputs and outputs (ex. withdrawals) in the Berrechid aquifer were incorporated into the model for each period. Calibration

a

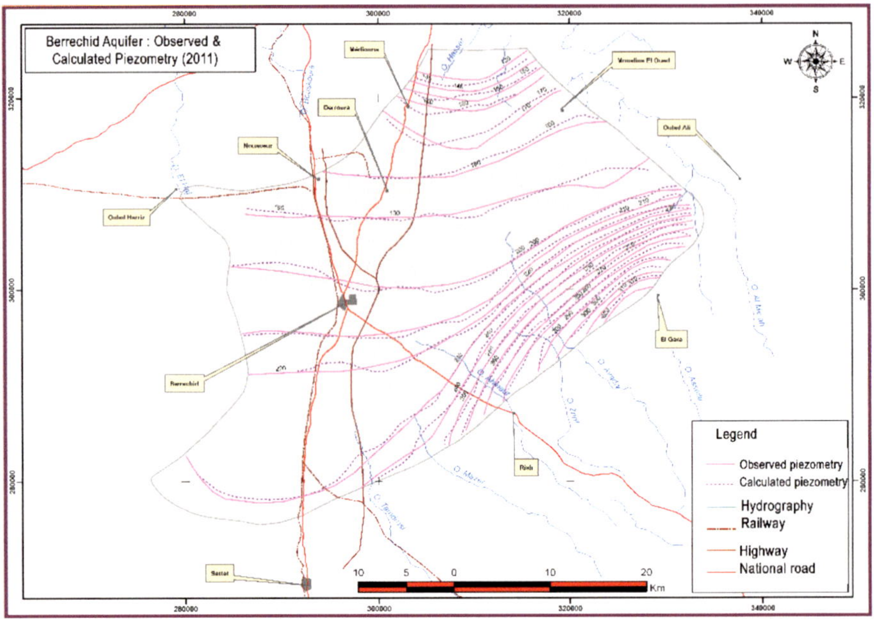

b

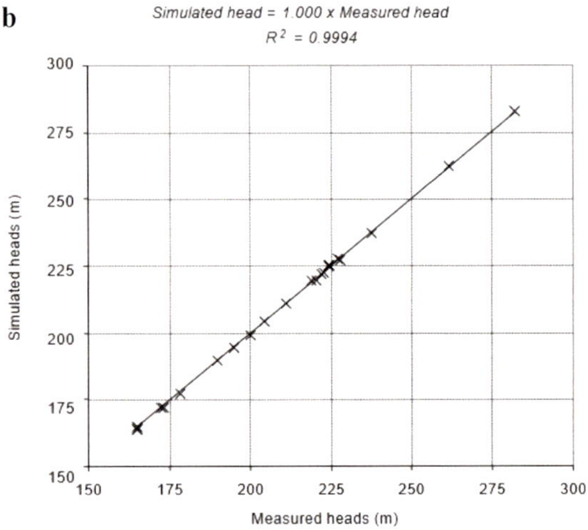

Fig. 4 **a** Comparison between observed and calculated piezometric levels for the reference state (1980). **b** Plot of simulated heads versus measured heads in the selected monitoring wells

Table 1 Groundwater balance calculated under steady-state conditions

Input	Q (m^3/year)	Output	Q (m^3/year)
– Rain infiltration	14.01	– Groundwater pumping	36.58
– Runoff infiltration from wadis	7.19	– Natural flow to Chaouia	7.31
– Natural flow by Settat plateau	24.83	– Drainage to Mellah wadi	2.17
Total IN	46.03	Total OUT	46.06
Input–Output			**0.03**
Error %			**0.004**

was performed using piezometric data collected from available observation wells selected for their long-term monitoring records under time series.

Figure 5a illustrates the calculated piezometric time series alongside measured data during the simulation period. The overall calculated piezometry at control observation wells demonstrates satisfactory agreement with the measured data.

The distribution of storage coefficient resulting from the model calibration indicates:

- Storage coefficient values corresponding to the effective porosity, where the aquifer is unconfined, range between 2.5 and 20%. This large variation is attributed to lateral facies changes within the aquifer.
- In areas where the aquifer is confined, the storage coefficient varies between 0.025 and 0.5%.

Reserve Evolution: The piezometric variation map (Fig. 5b) illustrates an overall spatial groundwater drawdown at the simulation end of the transient model, reflecting the depletion of groundwater reserves primarily caused by drought and extensive pumping, especially at the center of the aquifer. Therefore, the model allowed us to calculate different components of the aquifer groundwater balance, revealing a deficiency drawdown rate with an average of 2 m per year.

Several management scenarios were examined up to the year 2023 to explore options for managing the aquifer:

Management Scenario 1 (Fig. 6a): This scenario assumes a "liberal" exploitation without any actions or measures regarding withdrawals or control and conservation of water resources in the aquifer system. It represents uncontrolled overexploitation and intensive groundwater abstraction, leading to further declines in groundwater levels or even aquifer dewatering, particularly affecting the Pliocene horizon areas. The main model results show that extensive use of groundwater would deplete more than 80% of current reserves of the aquifer by the year 2035.

Management Scenario 2 (Fig. 6b): This scenario involves actions such as maintaining the current pumping volumes constant, even in the future years. The model results show that the aquifer declines continue, however with moderate depletion compared to the first scenario, eventually drawdowns reaching 8 m in the central part of the aquifer. In this case, approximately half of the reserves would be depleted by

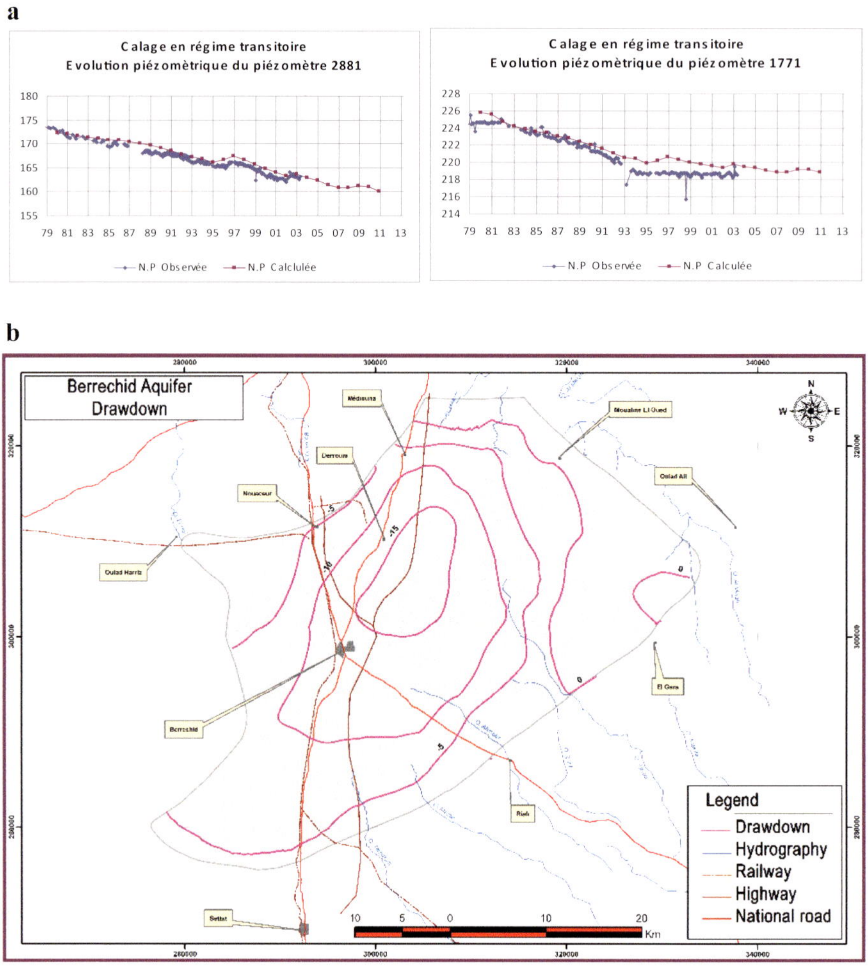

Fig. 5 **a** Comparison between observed and measured piezometric records. **b** Evolution of groundwater reserves in the aquifer

the year 2035. Despite evolving into a state of obvious overexploitation, the impacts of this scenario are less severe than in the uncontrolled exploitation scenario.

Hence, the incidents and outcomes of these groundwater management scenarios underscore the importance of developing a tool to control and manage groundwater withdrawals, involving direct actions and measures at the user level to promote groundwater protection and sustainability.

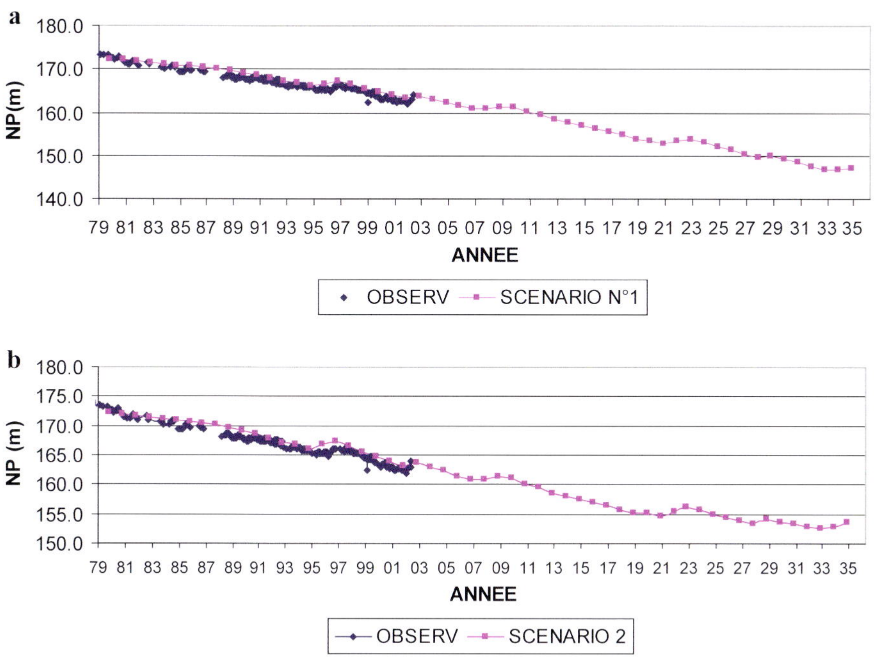

Fig. 6 Predicted piezometric evolution—**a** Scenario 1. **b** Scenario 2, in observation well 2881/20; X = 303,100; Y = 315,300; Z = 180

6 Decision Support System Using ArcGIS-MODFLOW Coupling

The droughts experienced by the country (covering also the study area) in recent years had a significant impact on the population's water supply, agriculture, energy production, and the natural environment. However, these extreme events have underscored the necessity of developing a decision support system (DSS) tool for assisting the policy makers and managing groundwater resource withdrawals in the Berrechid aquifer. This DSS tool is simply based on coupling GIS and MODFLOW for the designed regional model discussed in the previous sections.

The groundwater withdrawal management tool was developed for the Berrechid aquifer. It consists of an ArcGIS-Modflow coupling (interface), enabling to test the influence of pumping withdrawals either from domestic, industrial, or agricultural purposes. It also offers several advantages:

- Easy-to-use functionality without requiring expertise in modeling;
- Management of instantaneous withdrawals in any selected point or area.
- Rapid response to policy makers' needs regarding groundwater resource management.
- Calculation of drawdown due to withdrawals by point or area.

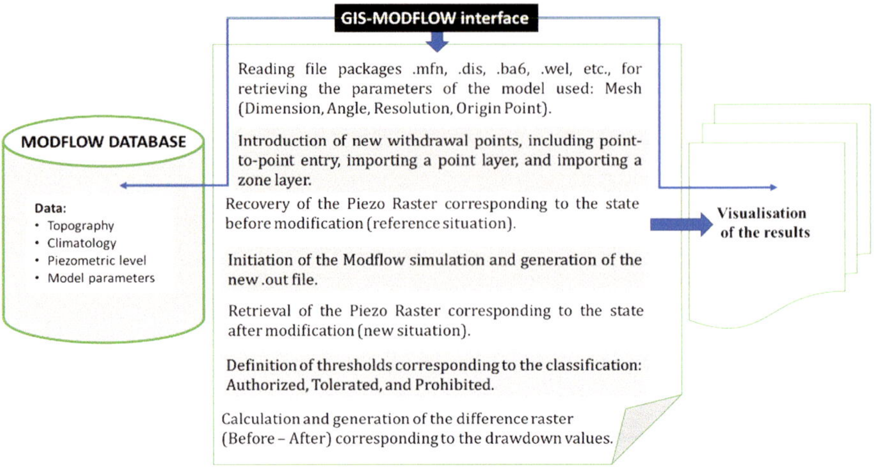

Fig. 7 Flowchart of the groundwater withdrawal management tool

- Definition of thresholds (Permitted, Tolerated, Prohibited) based on hydrogeological characteristics of each aquifer.
- Assessment of the impact of artificial aquifer recharge by point or area.

The development of this groundwater withdrawal management tool (DSS) aims to gather various functions into a common, user-friendly interface for an integrated information and decision support system (DSS). This system is designed to enable visualization and integration of databases, extraction of data for defined modeling projects, project management, simulation of prospective scenarios, and visualization of results in cartographic form (see Fig. 7).

The design of the application interfaces adheres to Windows standards, including input screens, drop-down menus, and direct selection of objects displayed on the screen by designation with the mouse. GIS functionalities are integrated to improve spatial data management and to automate and streamline the preparation of input variables. Significant effort has been dedicated to developing functions that facilitate consultation tailored to modeling needs.

The development of this groundwater withdrawal management tool (DSS) involves the following steps:

- Designing the ArcGIS-GIS-MODFLOW interface using ArcObjects libraries for spatial processing and Visual Studio 2010 (VB.Net) to manage Modflow data inputs and outputs (see Fig. 8).
- Reading various files (e.g., Mfn, dis, Ba6, Wel) to retrieve model parameters such as mesh details (size, angle, resolution, point of origin).

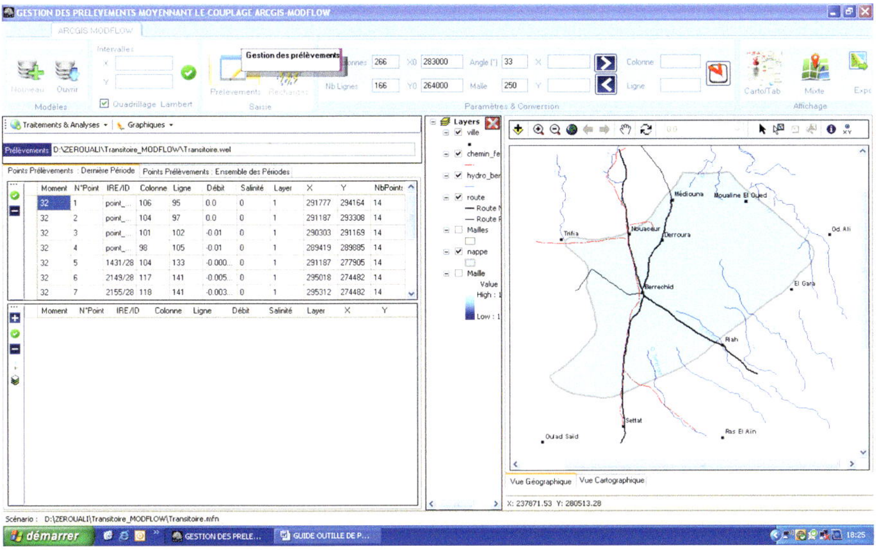

Fig. 8 GIS-MODFLOW interface development

- Introducing new withdrawal points (single or multiple) and importing point or area layers.
- Retrieving baseline piezometric raster data.
- Saving changes made to the withdrawal file (.Wel).
- Initiating Modflow simulation and generating new output files (.Out) (see Fig. 9).
- Retrieving piezometric raster data after modification (new scenario).
- Defining thresholds (Permitted, Tolerated, Prohibited).
- Calculating and generating a raster of drawdown differences (Before–After) (see Fig. 10).
- Classifying raster drawdown values based on user-defined classes.
- Presenting results (see Fig. 11).

7 Conclusions

The primary objective of this study was to develop a groundwater withdrawal management tool using GIS-MODFLOW coupling to enhance the management of groundwater resources. To do this, a comprehensive hydrogeological synthesis of the aquifer was conducted based on all available data.

The primary aquifer is located beneath the Quaternary silts, comprising Pliocene sandstone. In certain areas, the sandstone formations rise, and the aquifer extends horizontally into older horizons. The average thickness of the aquifer is approximately 20 m and can exceed 30 m alongside fossil channels of the pre-Pliocene hydrographic network.

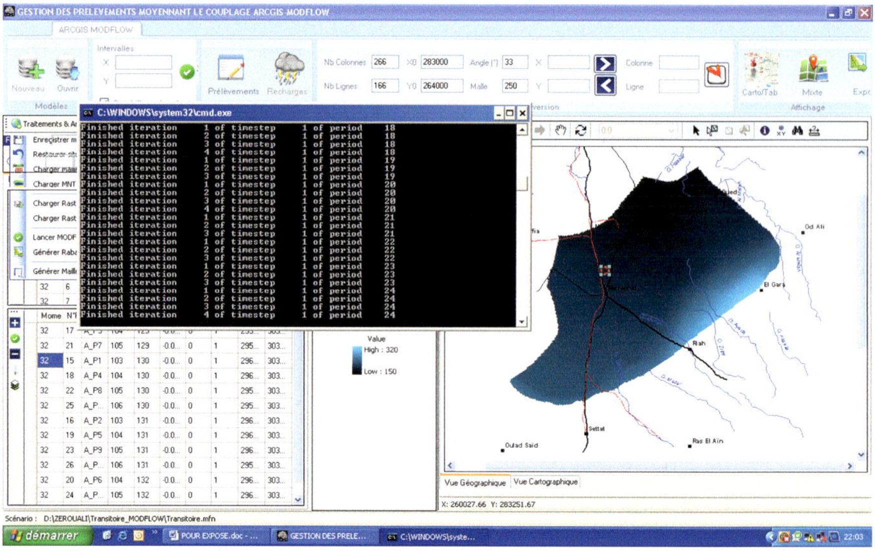

Fig. 9 Modflow simulation launch and generation of the new output file

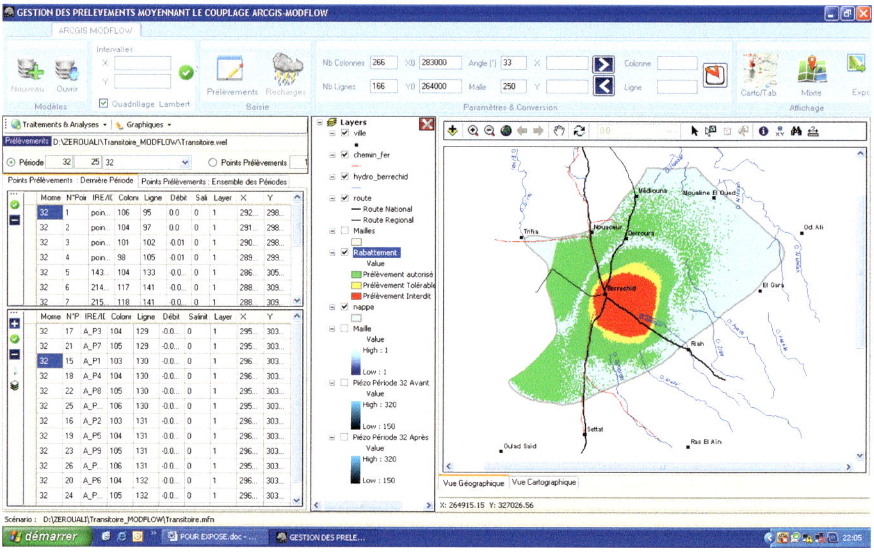

Fig. 10 Calculation and generation of raster for the difference (before–after) corresponding to the values of drawdown

The piezometric evolution of the aquifer indicates an equilibrium state from the 1950s until 1980, followed by a deficit period extending from 1981 to the present. This deficit period is characterized by continuous and sustained declines in the aquifer,

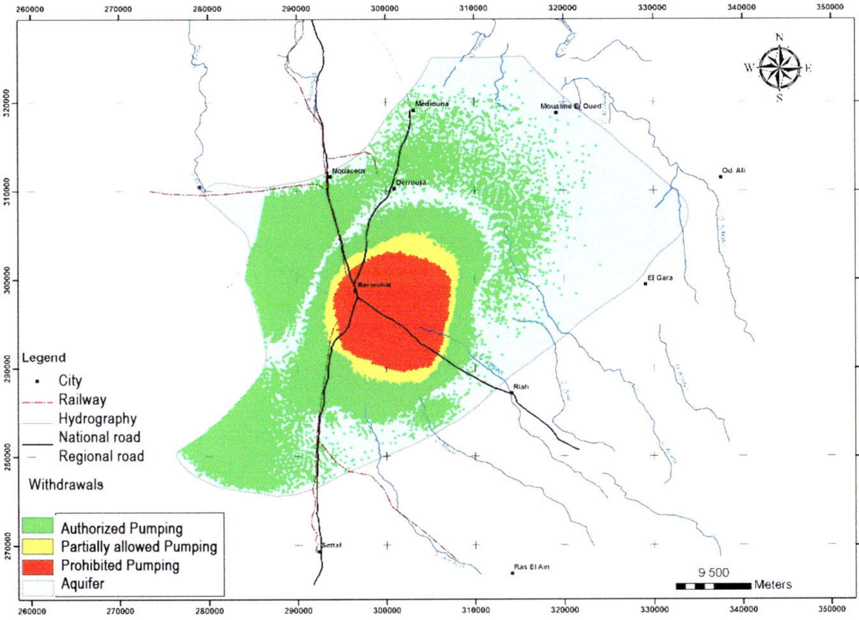

Fig. 11 Presentation of results under a vulnerability map showing sectors where withdrawals are prohibited (Red), partially allowed (Yellow) or authorized (Green)

attributed to a combination of reduced rainfall and increased withdrawals, particularly during drought episodes.

Hydrodynamic modeling of the Berrechid aquifer was conducted using the MODFLOW digital simulation software within the GMS pre-post-processor. The domain discretization consisted of 266 columns and 166 lines, delineating 44,156 regular square meshes with a side length of 250 m, including 24,401 active meshes. The mesh size was selected based on the density and variability of available data.

The reference piezometric state used for steady-state model calibration was the 1980 piezometric map, marking the onset of groundwater drawdown. The permeability distribution obtained after calibration revealed zones of very low K-values, primarily located south of the plain. These zones, characterized by values lower than 10^{-4} m/s corresponded to areas with the least transmissivity, where the groundwater circulated through thin sandstone fringes. The calibration of the steady-state model facilitated the reconstruction of the overall groundwater balance of the aquifer through the evaluation of its various components.

Transient modeling represents a continuation of the steady-state calibration process. The primary objective of this stage of hydrodynamic modeling is to calibrate the storage coefficient and homogenize aquifer system data under transient conditions. The selected simulation period spans 31 years, subdivided into annual intervals. Inputs and withdrawals in the Berrechid aquifer were incorporated into the model for each period, and calibration relied on piezometric data from long-time series observation points in the existing monitoring network of observation wells.

Overall, the calculated piezometric values closely match those measured at control observation wells, indicating satisfactory calibration results. The global piezometric variation map illustrates a general decline in the water table, reflecting reduced reserves attributed to drought and pumping activities. This requires the necessity of developing a withdrawal management tool to act directly at the user level and promote groundwater protection against overexploitation. The groundwater withdrawal management tool developed for the Berrechid aquifer utilizes ArcGIS-Modflow coupling, facilitating the calculation of drawdown resulting from withdrawals at specific points or zones and its spatial representation under a decision cartographic map of drawdowns showing the sectors where withdrawals are authorized or not in the study area.

Rational management of the Berrechid aquifer can be achieved through a collaborative agreement involving all water users and stakeholders in this hydrogeological entity. This kind of contract aims to:

- Encourage users to establish associations and cooperatives for collective and rational management of groundwater resources.
- Promote the cultivation of crops that require minimal water consumption to maximize the value of each cubic meter of water.
- Update the inventory of groundwater withdrawers and strengthen monitoring by enforcement teams.
- Reevaluate the current aquifer exploitation practices in other sectors by organizing farmers into associations.
- Review the existing system for processing water withdrawal permits, ensuring scrutiny regardless of the withdrawal rate.
- Enhance irrigation techniques by promoting water-saving systems as alternatives to the commonly used gravity-fed system.
- Implement a withdrawal management tool utilizing GIS-MODFLOW coupling to optimize the management of available groundwater resources.

References

ABHBC (2004) Hydrogeological study of the Berrechid aquifer, unpublished report from the Bouregreg and Chaouia hydraulic basin agency

ABHBC (2016) Bouregreg and Chaouia Hydraulic Basin Agency, Summary of previous studies and update of hydrogeological data from the Berrechid aquifer

DRPE (1997) Directorate of research and planning, study of the aquifer of the Berrechid plain, mission 1: description and analysis. 56p

El Mansouri B (1993) Structure and quantitative modeling of the Berrechid aquifer (Morocco). Validation using the geostatistical approach, quantitative hydrogeology, University of Sciences and Technologies of Lille, 241p

Macdonald MG, Harboaugh AW (1988) A modular three-dimensional finite difference groundwater flow model

MTPC (1977) Ministry of Public Works and Communications, Hydraulics Directorate, Water Resources Division, Water resources of Morocco, Volume 3, Edition of the geological service of Morocco

Water resources of Morocco (Volume 2) (1975) Plains and basins of Morocco, Atlantic, Note, Geological Service of Morocco

Integration of Water Quality Index (WQI) and Geographic Information Systems (GIS) to Assess Springs Water Quality in the El Hajeb-Meknes Area, Morocco

Abdennabi Alitane, Ali Essahlaoui, Ann Van Griensven, Steven Eisenreich, Narjisse Essahlaoui, Abdallah Elaaraj, Amina Kassou, Abdelouahed Essaied, and Yassine El Yousfi

Abstract The present research is focused on the evaluation of springs water suitability for human consumption based on water quality parameters. Because of the rising level of human activity, pollutants, especially trace elements, are entering

---

A. Alitane (✉) · A. Essahlaoui · N. Essahlaoui
Geoengineering and Environment Laboratory, Research Group "Water Sciences and Environment Engineering", Geology Department, Faculty of Sciences, Moulay Ismail University, Presidency, Marjane 2, BP 298, Meknes, Morocco
e-mail: abdennabi.alitane@edu.umi.ac.ma

A. Alitane · A. Van Griensven · S. Eisenreich
Water and Climate Department, Vrije Universiteit Brussels (VUB), 1050 Brussels, Belgium

A. Van Griensven
Water Resources and Ecosystems Department, IHE Delft Institute for Water Education, Delft, The Netherlands

A. Elaaraj
Natural Resources and Environment Laboratory, Geology Department, Polydisciplinary Faculty of Taza, Route d'Oujda, 1223, 35000 Taza, Morocco

Engineering Sciences and Techniques Center, Environment Department, Faculty of Science and Technology, Sidi Mohamed Ben Abdellah University of FEZ, 2202, 30000 Fes, Morocco

A. Kassou
Geo-Resource, Geo-Environment and Oasis Geological Heritage Research Group, Engineering Sciences and Techniques Laboratory, Geosciences Department, Faculty of Sciences and Techniques, Moulay Ismail University, Presidency, Marjane 2, BP 298, Meknes, Morocco

A. Essaied
Geoexploration and Geotechnics Research Team, Department of Geology, Faculty of Sciences, Moulay Ismail University of Meknes, Zitoune, 11201, Meknes, Morocco

Y. El Yousfi
GEE, Laboratory of Applied Sciences (LSA), ENSAH Al Hoceima, Abdelmalek Essaadi University, Tetouan, Morocco

N. Rebai et al. (eds.), *GIS, Applied Computing and Data Science for Water Management*, Lecture Notes in Geoinformation and Cartography,
https://doi.org/10.1007/978-3-031-63038-5_9

the aquatic system, posing harm to humans. The objectives of this study is: (i) to analyse the water quality variables, (ii) to compare their impact on water quality status of natural springs, and (iii) to display the Water Quality Index (WQI) results using Geographical Information System (GIS). The concentrations of water variables for investigated springs were analyzed adopting the Rodier standard method in the field using Multiparameter and in lab using a visible spectrophotometer and ICP-MS (ICPE-9000).The measured physico-chemical parameters were analyzed using a principal component analysis (PCA). A total of 24 water samples were taken from 12 springs to evaluate their suitability for human consumption. From the analysis of principal components (PCA), 73.87%. The variation was explained by the first three main factors. The variables pH, TDS, HCO_3^-, Na^+, Ca^{2+}, Mg^{2+} and Zn showed a strong correlation with F1. Following this, NO_3^-, Pb, Mn and Cu showed a strong correlation with F2, while $\mathbf{SO_4^-}$, CO, B, Ni and Cr showed a notable correlation with F3. The water quality index (WQI) was ranged from 49.66 to 95.10, averaging 76.52. About 17% of water samples were classified as excellent quality (WQI < 50), while the remaining 83% fell under the classification of good quality (50 < WQI < 100). This suggests that all the natural spring waters examined are suitable for human consumption.

Keywords WQI · SIG · Physico-chemical parameters · Springs water · Hajeb-Meknes area

1 Introduction

Water resources are some of the most important natural resources, they are found in many parts of the planet such as rivers, groundwater, springs, ponds, lakes, ocean and reservoirs (Vörösmarty et al. 2010; Alitane et al. 2022b), Water resource availability and quality are crucial for ecological health, sustainable development, and human survival (Xiao et al. 2019; Long and Luo 2020). Groundwater and springs are the main. The water sources utilized by the people of Morocco suffer from compromised quality due to a range of pollutant emissions, including those stemming from domestic, industrial wastewater, and agricultural (Choukr-Allah 2011). Various analyses, including physical, chemical, and biological analyses, can be utilized as references to assess the suitability of water for different purposes (Nollet and De Gelder 2000). Certainly, water quality is highly sensitive to climate change, as well as the flow of water within geological formations such as facies or faults, and anthropogenic activities (Rahman et al. 2014; Xiao et al. 2019; Elaaraj et al. 2023). There is currently growing concern about the quantity and diversity of toxins present in the environment. Trace elements, in particular, play a significant role in aquatic pollution and have garnered considerable attention worldwide (Kolarova and Napiórkowski 2021). Represent the most significant hazard due to their environmental toxicity, persistence, and carcinogenic potential (Khan et al. 2013; Bhuiyan et al. 2016). In the atmosphere, hydrosphere, lithosphere, and biosphere, trace elements are naturally distributed, but anthropogenic activities intensify their prevalence through the overuse of resources and improper waste disposal (Rathoure 2020). The presence of trace metals in various forms, such as soluble, exchangeable, included in minerals, complexed, or precipitated, determines their pollutant potential and the extent of their impact on water quality (Otmani et al. 2023). A number of these metals are non-degradable, persistent and toxic (Zeng et al. 2015), which would have adverse health and environmental impacts, and their accumulation in water (Guo et al. 2014; Turdi and Yang 2016). The initial suitability and quality of water for human consumption are compromised due to increased concentrations of trace elements. Using the WQI, five distinct classifications were established as follows: excellent water quality; good water quality; poor water quality; very poor water quality; deemed undrinkable water (Tong et al. 2021; El Yousfi et al. 2023b). The aim of this research project is to evaluate the physicochemical characteristics and trace elements of the drinking water springs within a sub-watershed located in Ouislane by using WQI and mapping the spatial distribution of water quality status of investigated sites by integrating the water quality model to a GIS environmental (El Yousfi et al. 2023a).

2 Material and Methods

2.1 Study Area

The Ouislane sub-watershed extends approximately 7.5 km from west to east and 51 km from north to south, covering an area of about 315.390 km^2. It is a part of R'Dom watershed (Alitane et al. 2022a), which covers two communes; Meknes in the north and El Hajeb in the south and the main source of water for irrigation and drinking for the population. Concerning the topography of the chosen area, elevation values range from 1,434 m (south) to 5 m (north). The Ouislane river is among the rivers of the R'Dom region (Alitane et al. 2022a), flows from south to north, receiving the discharge from major water springs. Twelve spring stations were chosen to evaluate the evolution of their quality. These sampling points are placed from upstream to downstream (Fig. 1). The main economic activity for the population is agriculture, which is dominated by crops of rainfed cereals, onion, and olive trees, with a small forest area in the southern part. About 22% of the total area is irrigated by water resources of the region (Sandu et al. 2010). In the Ouislane sub-watershed, the irrigation water sources can be surface water from rivers and groundwater extracted from springs or by using wells.

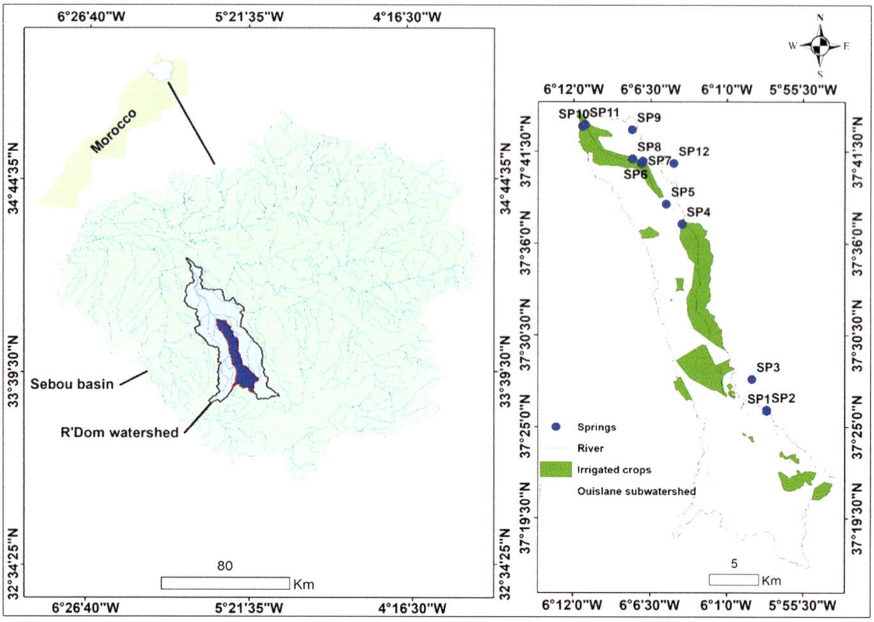

Fig. 1 Study area location

2.2 Sampling

The selection of water points was aimed at obtaining a complete picture of spring water quality. A total of twenty-four samples were taken for physicochemical and trace element analyses. The sampling water was filtered ($\varphi = 0.45\ \mu m$) into 60 ml glass bottles to complete the analysis under ICP-MS (ICPE-9000) (Fig. 2).

2.3 Methods

To analyses the hydro-geochemistry of the spring water, twelve samples were taken from visited stations for the period of May 2022. Water quality concerns had intensified in recent decades due to population growth, industrial activities and agricultural practices, as water resource is increasingly susceptible to pollution.

Water Parameters Analyzed

In this study, water samples were collected from spring sites and subjected to analysis in the laboratory. The analysis was conducted using a visible spectrophotometer and ICP-MS technique. The findings were used to compute the water quality index, which determines the appropriateness of the water for human consumption. PCA

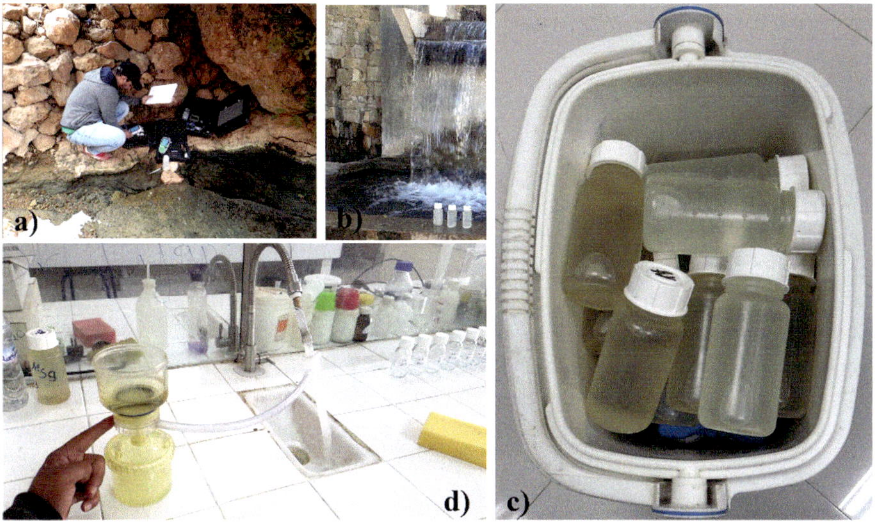

Fig. 2 Water analysis: **a** water parameters measurement, **b** and **c** water sampling, **d** water filtering

condensed the data into a reduced number of variables by employing weighted linear combinations of all these variables.

Statistical Tool

Statistical analysis was performed using SPSS (Statistical Package for the Social Sciences) version 23.0. Principal components analysis (PCA) was performed on all variables, and a correlation matrix between different variables was generated.

Water Quality Index

The Water Quality Index (WQI) is a mathematical expression using to representing water quality and its potential for use for a given activity. The Water Quality Index (WQI) assesses the drinkability of water based on the analyzed parameters. It is computed using the following equation (Guo et al. 2014):

$$WQI = \sum[W_i \times (\frac{C_i}{S_i}) \times 100]$$

where

Wi = wi/Σwi represents the relative weight,
wi is the weight attributed to the target element based on its relative apperceptive effects on human health and significance in terms of potability,
Ci represents the measured trace element concentrations,
Si is the Moroccan standard (NM) for drinking water (Meng et al. 2016).

3 Results and Discussion

3.1 Descriptive of Water Parameters Analyzed

The Table 1 summarize the water quality parameters of the investigated spring stations. The pH was ranged from 6.9 to 7.5, TDS was varied between 285 and 1001 mg/l, HCO_3^- was ranged from 113.5 to 288.1 mg/l, NO_3^- was ranged from 16 to 66 mg/l and SO_4^- was ranged from 3.4 to 27.6 mg/l. Regarding the cations concentrations, the Ca^{2+} was ranged from 37.6 to 65.6 mg/l, the Mg^{2+} was ranged between 11.6 and 42.3 mg/l and Na^+ was varied from 3.5 to 69.9 mg/l. The boron levels measured in the analysis was varied from 0.01 to 0.13 mg/l with an average of 0.049 mg/l. Chromium content was ranged between 0.05 and 0.08 mg/l. Chromium emerges as one of the biggest predominant elements detected in water samples, exceeding the upper limit of 0.05 g/l gives by the Moroccan norms or permissible levels in certain surveyed stations. Chromium, particularly in its hexavalent form, can cause health issues like stomach upsets, kidney, ulcers and liver damage, and even death (Strachan 2010). Lead and copper concentrations were found to be the lowest among all elements at all stations, measuring less than 0.02 mg/l. On the other hand, Nickel emerged as the most abundant element in all waters, with concentrations ranged from 0.027 to 0.125 mg/l. The measured values surpassed the maximum permissible limits established by the Moroccan standard, which is set at 0.02 mg/l. Manganese concentrations were only measured at stations (S1, S2, and S10) and were absent at the other stations. In general, the differences among water samples were more noticeable, particularly in the instances of boron, chromium, and nickel.

For this study, the Moroccan standard (NM) and the World Health Organization (WHO) drinking water standards were selected. Table 2 presented the statistically analyzed parameters alongside the maximum acceptable standard values. The Cr in mineral waters recommended by Moroccan standard and WHO is around about 0.05 mg/l, which was exceed in the majority of sampled waters. The presence of chromium (Cr), particularly in its hexavalent form, is linked to cancer when detected in drinking water, which is a major public health issue (Georgaki et al. 2023). The table below shows the water quality variables compared to Moroccan standard (NM 03.7.001) and World Health Organization (WHO 2011).

Table 1 Water parameters concentrations in the investigated stations

Stations	pH	TDS	HCO_3^-	NO_3^-	SO_4^-	Ca^{2+}	Mg^{2+}	Na^+	B	Co	Cr	Cu	Mn	Ni	Pb	Zn	WQI
Sp1	7.46	285	206.77	58	11.5	51.6	42.3	4.19	0.01	0.054	0.08	0.02	0.03	0.09	0.01	0.02	**95.1**
Sp2	7.3	361	288.13	30	6.8	48.3	29.5	6.69	0.03	0.056	0.08	0	0.04	0.09	0	0.03	**79.16**
Sp3	7.33	291	263.54	16	3.4	43.2	36.1	3.47	0.02	0.055	0.08	0.01	0	0.09	0	0.03	**66.13**
Sp4	7.45	376	153.65	17	5.7	37.6	31.5	20	0.02	0.052	0.05	0	0	0.03	0	0.03	**49.72**
Sp5	7.44	492	160.16	18	3.8	42.8	19.2	35.1	0.02	0.052	0.05	0	0	0.03	0	0.03	**49.66**
Sp6	7.28	504	238.97	49	18.1	67.5	28.4	22	0.05	0.05	0.06	0	0	0.09	0	0.04	**84.04**
Sp7	7.14	486	148.12	52	14.3	78.9	15.1	20	0.05	0.05	0.06	0	0	0.09	0	0.04	**79.35**
Sp8	6.9	615	148.812	49	21.2	65.6	25.8	51.6	0.08	0.05	0.07	0	0	0.10	0	0.04	**81.99**
Sp9	7.34	464	113.54	61	9.7	72.4	11.6	7.93	0.02	0.04	0.06	0	0	0.09	0	0.03	**77.67**
Sp10	7.05	1001	210.16	44	23.3	85.6	28.4	68.4	0.12	0.05	0.07	0.003	0.013	0.13	0	0.03	**91.66**
Sp11	7.06	858	248.66	18	27.6	69.7	26.3	69.9	0.13	0.05	0.08	0.003	0	0.12	0	0.03	**78.14**
Sp12	7.29	321	163.54	66	13.8	63.9	23.5	9.11	0.03	0.05	0.06	0.005	0	0.09	0	0.05	**85.58**

Table 2 Statistics of water variables (mg/l)

Par	Min	Max	Avg	SD	NM	WHO
pH	6.9	7.460	7.253	0.17	8.4	8.5
TDS	285	1001	504	214	1000	1000
HCO_3^-	113	288	195	52	–	120
NO_3^-	16	66	39.8	18.1	50	50
SO_4^-	3.4	27.6	13.3	7.61	400	400
Ca^{2+}	37.6	85.8	60.6	14.9	–	150
Mg^{2+}	11.6	42.3	24.5	8.2	–	30
Na^+	3.5	69.9	26.5	23.3	–	200
B	0.012	0.134	0.049	0.04	0.5	0.5
Co	0.048	0.058	0.054	0.01	50	50
Cr	0.035	0.082	0.066	0.01	0.05	0.05
Cu	0	0.015	0.003	0.01	2	2
Mn	0	0.04	0.007	0.01	0.5	0.04
Ni	0.062	0.125	0.096	0.015	0.02	0.07
Pb	0	0.006	0.001	0.002	0.01	0.01
Zn	0.023	0.047	0.034	0.01	3	3

3.2 Correlation Matrix Analysis

In this research, the Pearson correlation matrix was generated to explore potential sources, origins, and correlations of trace elements in the source samples (El Yousfi et al. 2022). A correlation matrix elucidates the relationships between trace elements, unveiling new associations among them. The water quality parameters demonstrate significant positive correlations (R2 > 0.5) and negative correlations (−0.5 > R2). Additionally, they exhibit weak positive correlations (R2 < 0.5) and negative correlations (R2 > −0.5), as well as correlations that are insignificant when the correlation coefficient (r) is close to zero (Table 3).

Table 3 shows a strong correlation between Ca^{2+} and Mg^{2+} ($R^2 = 0.91$), this comes when water interacts with limestone (composed mainly of calcium carbonate, $CaCO_3$) and dolomite (containing both calcium carbonate and magnesium carbonate, $CaMg(CO_3)_2$), a process known as dissolution occurs. In this chemical reaction, carbonic acid, which forms when water combines with carbon dioxide in the atmosphere, reacts with the carbonate minerals in limestone and dolomite. The result is the release of calcium (Ca^{2+}) and magnesium (Mg^{2+}) ions into the water. These dissolved ions contribute to the water's hardness, influencing its mineral content. This natural process affects the composition of spring waters in the rich El Hajeb-Meknes regions, which are rich in limestone and dolomite. There were also a good correlation between Co and Cr ($R^2 = 78$), B and Ni, which are suggest that changes in one variable are strongly associated with changes in the other, and the relationship is close

Table 3 Pearson correlation matrix for hydrochemical variables of springs

Par	pH	TDS	HCO_3^-	NO_3^-	SO_4^-	Ca^{2+}	Mg^{2+}	Na^+	B	Co	Cr	Cu	Mn	Ni	Pb	Zn
pH	1															
TDS	**0.71**	1														
HCO_3^-	**0.53**	**0.60**	1													
NO_3^-	**−0.62**	**−0.68**	−0.46	1												
SO_4^-	0.02	0.46	0.12	−0.21	1											
Ca^{2+}	0.39	**0.51**	**0.56**	−0.15	**0.62**	1										
Mg^{2+}	0.25	**0.59**	**0.60**	−0.16	**0.62**	**0.91**	1									
Na^+	0.47	**0.70**	0.47	−0.29	0.23	**0.60**	**0.57**	1								
B	−0.16	0.08	0.13	0.05	0.53	0.46	0.31	0.45	1							
Co	−0.16	0.33	0.07	−0.31	**0.64**	0.14	0.19	0.21	**0.67**	1						
Cr	−0.26	0.19	−0.02	−0.33	**0.61**	−0.11	0.03	−0.27	0.22	**0.78**	1					
Cu	0.36	**0.54**	−0.08	**−0.63**	0.48	0.13	0.17	0.07	−0.24	0.27	0.41	1				
Mn	0.23	0.49	0.02	−0.41	0.31	−0.13	−0.03	0.03	−0.16	0.33	0.46	0.49	1			
Ni	−0.32	−0.07	−0.02	0.27	0.45	0.42	0.29	0.41	**0.95**	**0.57**	0.10	−0.32	−0.21	1		
Pb	0.18	0.48	−0.10	−0.42	0.34	0.00	0.10	0.14	−0.28	0.04	0.21	0.75	**0.58**	−0.36	1	
Zn	0.10	0.32	0.30	0.05	−0.22	0.20	0.39	**0.59**	0.09	0.07	−0.31	−0.21	−0.18	0.18	−0.25	1

to a perfect positive linear correlation. The reasons for a high correlation between the both elements might be due to similar geological or environmental sources or industrial activities that affect both elements concurrently.

3.3 Principal Components Analysis (PCA)

According to the PCA results of the 12 water samples (Fig. 3), the total water quality variance account about 73.87%, which was contributed on F1 (32.39%), F2 (22.67) and F3 (18.71%), respectively (Fig. 4). pH, TDS, HCO_3^-, Na^+, Ca^{2+}, Mg^{2+}, Zn showed a strong correlation with F1, then, NO_3^-, Pb, Mn and Cu have a high correlation to F2 and SO_4^-, CO, B, Ni and Cr displayed a good correlation with F3 (Table 4).

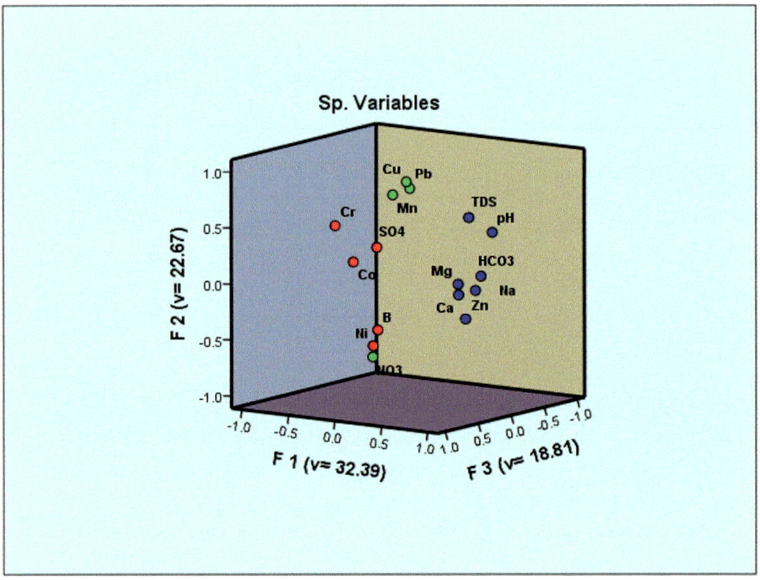

Fig. 3 Distribution of water quality parameters in PCA plan

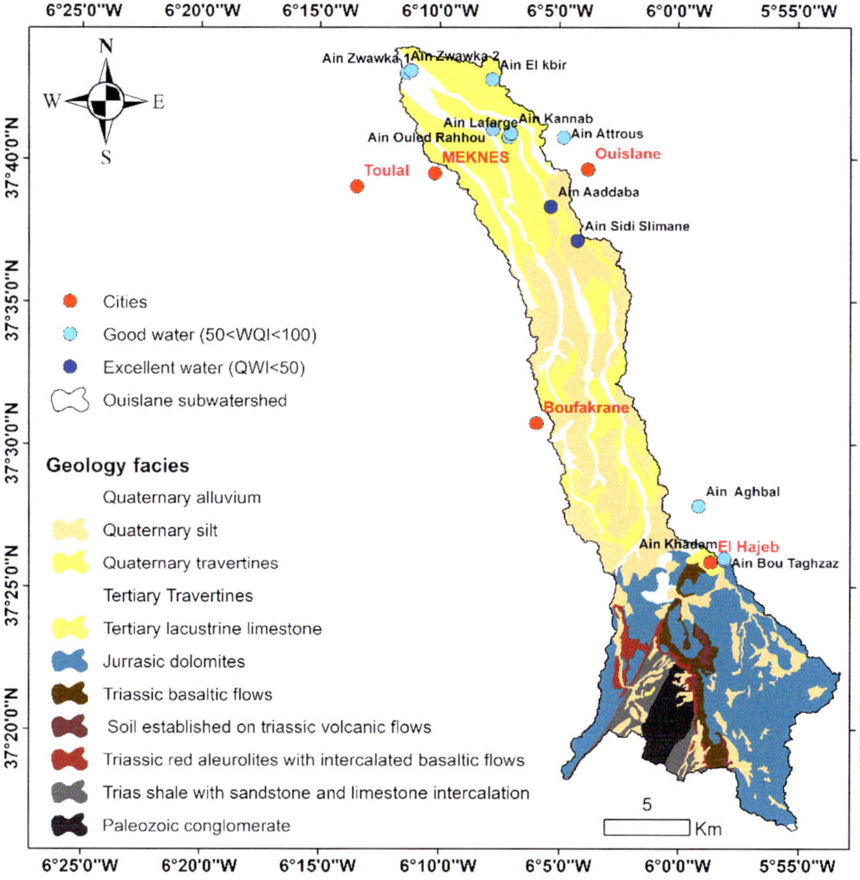

Fig. 4 Simplified geological map and spatial distribution of water quality categories

3.4 *Water Quality Index (WQI)*

Water quality index in the visited springs was varied between 49.7 and 95.1. Based on the Water Quality Index (WQI) classification and GIS environment (Fig. 4), the investigated stations were classified as having excellent water quality (WQI < 50), which was represent 17% from the total sampled water, and a good quality (50 < WQI < 100), which was represent 83% from the total sampled water. The consumption of contaminated water poses public health risks and is a serious problem in rural areas.

Table 4 Rotated component matrix

PCA		pH	TDS	HCO_3^-	NO_3^-	SO_4^-	Na^+	Ca^{2+}	Mg^{2+}	Pb	Mn	Zn	Co	B	Cu	Cr
Factor	F1	0.65	0.73	0.76	−0.35	0.22	0.85	0.80	0.78	−0.01	−0.08	0.57	0.02	0.27	0.04	−0.32
	F2	0.42	0.63	0.09	−0.73	0.41	0.00	−0.02	0.07	0.79	0.75	−0.32	0.27	−0.31	0.88	0.53
	F3	−0.37	0.10	−0.04	0.03	0.78	0.17	0.35	0.33	−0.06	0.11	−0.07	0.86	0.84	0.07	0.64

4 Conclusion

The results of the physico-chemical analysis of springs water in the El Hajeb-Meknes region, obtained during the period of May 2022, can be considered acceptable and present no danger for human consumption. Principal Component Analysis (PCA) of the physicochemical data was enabled to discern the prevailing correlations among the various parameters. The springs water come from shallow aquifer which is characterized by a diversity of geological facies. These could have been the result of several mechanisms responsible for the mineralization of the water. This decision was made in light of the potential negative impacts of these elements on water quality, which may include toxicity. However, despite these concerns, the water quality at the surveyed stations was categorized as ranging from good (50 < WQI < 100) to excellent (WQI < 50). So, the water quality monitoring illustrates that 83% of the controlled points with a good quality and 17% of excellent quality.

References

Alitane A, Essahlaoui A, El Hafyani M et al (2022a) Water erosion monitoring and prediction in response to the effects of climate change using RUSLE and SWAT equations: case of R'Dom watershed in Morocco. Land 11:93

Alitane A, Essahlaoui A, Van Griensven A et al (2022b) Towards a Decision-making approach of sustainable water resources management based on hydrological modeling: a case study in central Morocco. Sustainability 14:10848

Bhuiyan MAH, Bodrud-Doza M, Islam AT et al (2016) Assessment of groundwater quality of Lakshimpur district of Bangladesh using water quality indices, geostatistical methods, and multivariate analysis. Environ Earth Sci 75:1–23

Choukr-Allah R (2011) Comparative study between Moroccan water strategies and WFD. Dialogues on Mediterranean water challenges: rational water use, water price versus value and lessons learned from the European Water Framework Directive Bari: CIHEAM, pp 181–188

El Yousfi Y, Himi M, Aqnouy M et al (2023a) Pollution vulnerability of the Ghiss Nekkor Alluvial aquifer in Al-Hoceima (Morocco), using GIS-based DRASTIC model. Int J Environ Res Public Health 20:4992

El Yousfi Y, Himi M, El Ouarghi H et al (2023b) Assessment and prediction of the water quality index for the groundwater of the Ghiss-Nekkor (Al Hoceima, Northeastern Morocco). Sustainability 15:402

El Yousfi Y, Himi M, El Ouarghi H et al (2022) Hydrogeochemical and statistical approach to characterize groundwater salinity in the Ghiss-Nekkor coastal aquifers in the Al Hoceima province, Morocco. Groundwater Sustain Dev 19:100818

Elaaraj A, Lhachmi A, Tabyaoui H et al (2023) Highlighting volcanic outcrops by mapping geological lineaments using satellite data in the Saka region, north-east Morocco. Ecol Eng & Environ Technol

Georgaki M-N, Charalambous M, Kazakis N et al (2023) Chromium in water and carcinogenic human health risk. Environments 10:33

Guo H, Wen D, Liu Z et al (2014) A review of high arsenic groundwater in Mainland and Taiwan, China: distribution, characteristics and geochemical processes. Appl Geochem 41:196–217

Khan S, Shahnaz M, Jehan N et al (2013) Drinking water quality and human health risk in Charsadda district, Pakistan. J Clean Prod 60:93–101

Kolarova N, Napiórkowski P (2021) Trace elements in aquatic environment. Origin, distribution, assessment and toxicity effect for the aquatic biota. Ecohydrol Hydrobiol 21:655–668
Long J, Luo K (2020) Elements in surface and well water from the central North China Plain: enrichment patterns, origins, and health risk assessment. Environ Pollut 258:113725
Meng Q, Zhang J, Zhang Z, Wu T (2016) Geochemistry of dissolved trace elements and heavy metals in the Dan River Drainage (China): distribution, sources, and water quality assessment. Environ Sci Pollut Res 23:8091–8103
Nollet LM, De Gelder LS (2000) Handbook of water analysis. CRC Press
Otmani M, Echajia M, Alami L et al (2023) Trace metal elements in different categories of drinking water by exploratory analysis. Appl J Environ Eng Sci 9(2):69–81
Rahman MS, Saha N, Molla AH (2014) Potential ecological risk assessment of heavy metal contamination in sediment and water body around Dhaka export processing zone, Bangladesh. Environ Earth Sci 71:2293–2308
Rathoure AK (2020) Heavy metal pollution and its management: Bioremediation of heavy metal. In: Waste management: concepts, methodologies, tools, and applications. IGI Global, pp 1013–1036
Sandu M, Mosanu E, Gladchi V et al (2010) Study of springs water quality as sources of potable water and for irrigation in rezina district. Chem J Mold 5:84–89
Strachan S (2010) Trace elements. Curr Anaesth Crit Care 21:44–48
Tong S, Li H, Tudi M et al (2021) Comparison of characteristics, water quality and health risk assessment of trace elements in surface water and groundwater in China. Ecotoxicol Environ Saf 219:112283
Turdi M, Yang L (2016) Trace elements contamination and human health risk assessment in drinking water from the agricultural and pastoral areas of Bay County, Xinjiang, China. Int J Environ Res Public Health 13:938
Vörösmarty CJ, McIntyre PB, Gessner MO et al (2010) Global threats to human water security and river biodiversity. Nature 467:555–561
WHO (2011) Guidelines for drinking-water quality: World Health Organization. Distribution and Sales, Geneva, p 27
Xiao J, Wang L, Deng L, Jin Z (2019) Characteristics, sources, water quality and health risk assessment of trace elements in river water and well water in the Chinese Loess Plateau. Sci Total Environ 650:2004–2012
Zeng X, Liu Y, You S et al (2015) Spatial distribution, health risk assessment and statistical source identification of the trace elements in surface water from the Xiangjiang River, China. Environ Sci Pollut Res 22:9400–9412

Long-Term Assessment of Spatiotemporal Silting Patterns in Moroccan Dams via Bathymetry-GIS Integration

Said Mohafid, Laila Stour, Ayoub Benchara, and Ali Agoumi

Abstract In regions facing the dynamic challenges of environmental change, the sustainable development of water management strategies is paramount. In the Moroccan landscape, where the increase in water demand, driven by population growth and intensified by the adverse effects of climate change accentuating the siltation of reservoirs, the importance of dams in the management of water resources requires a thorough understanding of spatiotemporal siltation models. This research work presents an in-depth investigation of siltation phenomena within Moroccan dams, over a long period. Using an integrative methodology, the research combines century-old bathymetric data with Geographic Information System (GIS) technologies. The resulting interactive maps make it possible to identify regions most sensitive to dam siltation and provide a complete overview of temporal changes. The analysis highlights the omnipresence of siltation in all Moroccan watersheds. The Moulouya basin is distinguished by its unique pedological characteristics and its arid climate, exacerbated by intense rainfall and deforestation. The results provide valuable information for decision-makers and water resource managers, offering essential information for strategic planning and the implementation of targeted measures to mitigate the impact of siltation.

Keywords Siltation · Bathymetry · Geographic Information System (GIS) · Spatial and temporal distribution · Water resources management · Morocco

S. Mohafid (✉) · L. Stour
Process and Environment Engineering Laboratory, Faculty of Sciences and Techniques of Mohammedia, Hassan II University of Casablanca, Mohammedia, Morocco
e-mail: said.mohafid-etu@etu.univh2c.ma

A. Benchara
Directorate General of Hydraulics, Moroccan Ministry of Equipment and Water, Rabat, Morocco

A. Agoumi
Civil Engineering, Hydraulics, Environment, and Climate Change Laboratory, Hassania School of Public Works, Casablanca, Morocco

N. Rebai et al. (eds.), *GIS, Applied Computing and Data Science for Water Management*, Lecture Notes in Geoinformation and Cartography,
https://doi.org/10.1007/978-3-031-63038-5_10

1 Introduction

Water resource sustainability is a global concern, particularly in regions like Morocco that grapple with arid and semi-arid climates. Large dams, constructed over decades, stand as critical infrastructures to manage this precious resource. However, the persistent challenge of silting poses a threat to the efficiency and longevity of these reservoirs. Silting diminishes storage capacities, affecting water quality and impeding various essential functions such as irrigation and energy generation (Remini and Hallouche 2005; Lahlou 1994).

Understanding the intricate dynamics of spatiotemporal silting patterns is essential for effective water resource management. Previous global studies have shown that, on average, the siltation rate of dams worldwide is around 1%, with significant variations from one country to another, up to 3.27% in Tanzania (Walling 2012).

This study focuses on the Moroccan context, where the specific terrain and climate characteristics contribute to unique silting patterns.

The advent of GIS technologies and the availability of historical bathymetric data offer an unprecedented opportunity to delve into the long-term assessment of silting patterns. By integrating these datasets, we aim to uncover nuanced insights into how Moroccan dams have evolved in response to changing environmental conditions.

Our research focuses on conducting a comprehensive analysis of dam siltation in Morocco, spanning almost a century. This entails integrating historical bathymetric data and utilizing GIS technologies. We use bathymetric measurements to trace the trajectory of siltation in the context of water resource scarcity, further exacerbated by recent years of drought associated with climate change in the region (Masih et al. 2014; Yobom 2020).

The subsequent sections present our work and findings, which are integral to our broader research aiming to valorize sediments retained by dams by utilizing them as raw materials across various industries (Mohafid et al. 2021, 2023). This article complements existing qualitative analyses and quantitative trends, introducing a novel GIS application designed for the spatiotemporal representation of solid deposits.

Our GIS application allows us to map variations in the spatial distribution of sediments accumulated over time, providing a comprehensive and detailed insight into dam siltation. This innovative approach reveals the most affected areas and prominent periods of accumulation.

The outcomes of our research contribute to a deeper understanding of the factors influencing dam siltation and furnish crucial information for decision-makers and water resource managers. This data is imperative for the development of sustainable management strategies, aiming to mitigate the adverse effects of siltation and ensure the long-term preservation of these vital infrastructures.

The following sections of this article are organized as follows: The next section provides a detailed description of the methodology used for collecting bathymetric data and implementing GIS analysis. Following that, we present and discuss the results obtained, emphasizing significant spatiotemporal trends in the siltation of large dams. Finally, we conclude by underscoring the importance of the spatiotemporal approach in comprehending dam siltation and its implications for the sustainable management of water resources in Morocco.

2 Methods

2.1 Acquisition and Preprocessing of Bathymetric Data

The bathymetric data utilized in this research, obtained from the Moroccan Department of Water, spans almost a century. These datasets comprise a comprehensive compilation of bathymetric surveys conducted at various points in history, offering a unique perspective on the development of dam reservoirs. The managers of the reservoirs of the dams in Morocco use acoustic-sounding techniques to have precise surveys and depth profiles of the reservoirs.

Before integration into the analysis, the raw data underwent a thorough cleaning and standardization process. This involved assessing data quality, rectifying errors and inconsistencies, and converting it into a standardized format suitable for further processing in the GIS.

Following meticulous verification and correction, these measurements provide valuable insights into reservoir evolution. The dataset covers 60 out of the 152 large dams in Morocco, with usable data available for 53 of these dams. Collectively, these 53 dams represent an initial reservoir volume of 17.57 Mm^3, accounting for over 92% of the total 19.19 Mm^3 (MEE 2023).

2.2 Crafting the GIS Application for Spatiotemporal Analysis

We meticulously designed a tailored GIS application for this investigation to facilitate the streamlined analysis of bathymetric data. This application was constructed utilizing specialized programming libraries dedicated to GIS, incorporating a myriad of essential features:

1. **Data Integration**: The bathymetric data, already subjected to meticulous cleaning, has seamlessly been incorporated into the application. This integration enables their visualization and geospatial manipulation, as depicted in Fig. 1.
2. **Spatiotemporal Representation**: Our application facilitated the generation of dynamic maps and graphs, providing an interactive platform for observing the temporal and spatial variations in reservoir siltation. Users could explore depth maps, sediment profiles, and animated representations illustrating the evolution of siltation over different time intervals (Fig. 2).

Fig. 1 Geospatial Representation of Sediment Distribution within Dam Reservoirs in 2022

Fig. 2 The temporal evolution of siltation is exemplified for four dams

3. **Temporal Trend Analysis**: We have conducted in-depth trend analysis using techniques such as linear regression to discern patterns in sediment deposition over time (Fig. 3).

Fig. 3 Map illustrating the trends in dam siltation analysis

2.3 Computer Tools Used for the Development of GIS

To develop our GIS application, we utilized the open-source ***PHP programming language***, a versatile server-side scripting language known for building dynamic websites and interactive web applications (Apiag et al. 2023).

The dynamic web application and website were constructed using the open-source ***CodeIgniter Framework***, coded in PHP. CodeIgniter is renowned for its user-friendly nature, lightweight design, and straightforward functionality. It is a toolkit for facilitating the construction of websites using PHP. Its objective is to expedite project development compared to coding from the ground up, offering a comprehensive set of libraries for frequently required tasks. It features a straightforward interface and a logical structure to access these libraries, allowing developers to concentrate on their projects more creatively while minimizing the necessary code for specific tasks (BCT 2023).

For creating interactive maps and the web mapping application, we employed the capabilities of the ***OpenLayers Library***. This library offers a comprehensive suite of mapping and geolocation tools, enabling the development of interactive maps, geospatial data visualization, on-map element drawing, marker integration, and more.

OpenLayers simplifies the process of incorporating a dynamic map into any web page. It is capable of presenting map tiles, vector data, and markers retrieved from various sources. The development of OpenLayers aims to encourage the utilization of diverse geographical information (OSGF 2023).

By incorporating the ***Bing Maps*** online mapping service by Microsoft, we utilized its capabilities for mapping, navigation, and geolocation functionalities.

2.4 Evaluation of the Results

After conducting the GIS analysis, we actively evaluated the results to identify significant patterns in dam reservoir siltation. We focused on examining distribution maps, trend graphs, and spatial analyses to pinpoint regions most profoundly impacted by siltation and to understand temporal variations. The upcoming sections will provide additional details on these findings.

For user accessibility, the GIS application is available on the website https://moroccandams.net/.

3 Results

3.1 Temporal and Spatial Changes in Siltation Patterns

The GIS-generated maps provided a detailed insight into the complex and evolving patterns of siltation within Moroccan dam reservoirs. Our findings highlight significant changes in reservoir depths over the years, indicating the gradual accumulation of sediments. Through distribution maps and evolution curves, we highlight the regions most affected by siltation, offering a clear understanding of the prevailing trends in sediment deposition.

Notably, the Moulouya basin in north-eastern Morocco, especially near the estuary, emerges as the most severely impacted area, notably affecting the Mechraâ Homadi and Mohamed V dam reservoirs. In the central Oum Errbia watershed, the Bine El Ouidane and Ahmed El Hansali dams experience siltation rates exceeding five Mm^3 annually. In the Tensift basin, the Sidi Driss and Timinoutine dams are currently confronting near-total siltation.

As an illustrative case, we will conduct an in-depth analysis of the natural conditions within the Moulouya watershed in the subsequent sections.

3.2 Identifying Areas Prone to Vulnerability

The GIS application played a key role in identifying areas within dam reservoirs that are more prone to siltation. The maps generated through this analysis not only

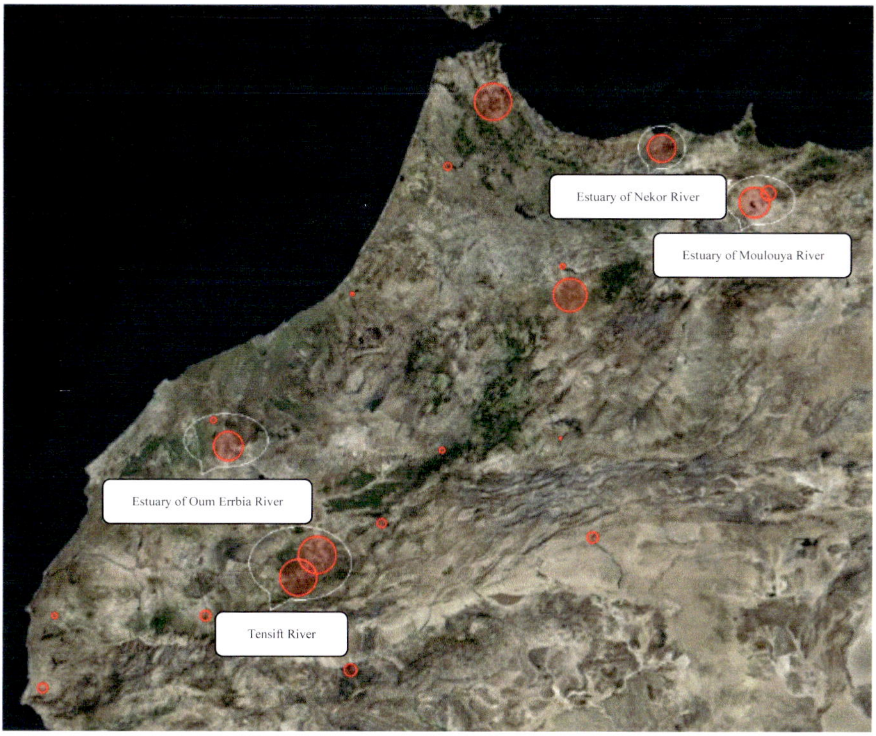

Fig. 4 Map showing the areas where the risk of siltation is high

revealed regions with an increased susceptibility to sediment accumulation but also provided valuable insights into the specific nature of these high-risk zones. Recognizing these vulnerable areas is vital for a comprehensive assessment of dam storage capacity, necessitating focused attention in future water resource management strategies. The estuaries of the Moulouya River, Nekor River, and Oum Errbia River, along with distinct sectors of the latter and the Tensift River, emerged as particularly susceptible zones, as illustrated in Fig. 4.

3.3 Temporal Trend Analysis

The analysis of temporal trends has revealed a variety of siltation patterns. Some dams consistently experience an increase in siltation over the decades, while others go through phases of rapid siltation followed by periods of stability. Understanding these temporal patterns is essential for developing strategies related to dam maintenance and management.

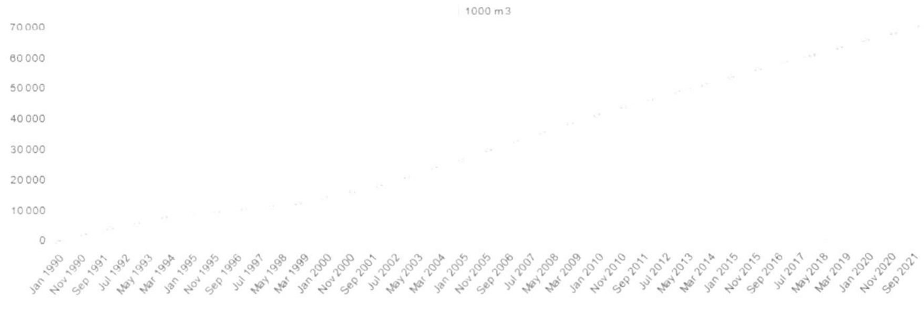

Fig. 5 Siltation evolution curve for the Allal Al Fassi Dam on the Sebou River

Figure 5 depicts the siltation evolution curve of the Allal Al Fassi dam located on the Sebou River.

3.4 Implications for Water Resources Management

The obtained results emphasize the critical importance of vigilant monitoring of siltation in dam reservoirs and the subsequent implementation of appropriate measures to counteract its repercussions. A comprehensive understanding of the complex spatiotemporal dynamics of siltation is the cornerstone of informed decision-making, particularly in the field of water resource management. This involves not only identifying high-risk areas but also devising strategic plans for siltation removal operations and establishing routine maintenance protocols to ensure the sustained functionality of the infrastructure. The detailed insights provided by the spatiotemporal analysis become integral tools for shaping proactive and effective strategies in the sustainable management of dam reservoirs.

3.5 Limitations

Acknowledging several limitations is crucial for a comprehensive understanding of this work. Firstly, historical data might carry discrepancies and inaccuracies, potentially influencing the outcomes of the analysis. Additionally, the spatiotemporal examination relies on one-off surveys, and the application of more advanced modeling techniques could provide improved precision in siltation estimates. These limitations underscore the need for continued refinement and advancement in data collection methods and analytical approaches for future studies in this domain.

4 Discussion

4.1 Analysis of Natural Conditions Explaining the Produced Maps—A Case Study of the Oum Errbia Watershed

Stretching over a length of 550 km, the Oum Errbia River drains its vast basin, which covers an area of approximately 35,000 km^2, constituting 5.1% of the total area of the country (Fig. 6).

The Oum Errbia watershed exhibits remarkable climatic diversity, with rainfall ranging from 300 to 1,100 mm. This diversity translates into distinct climatic zones, ranging from the coastal climate along the Atlantic coast to an orographic climate

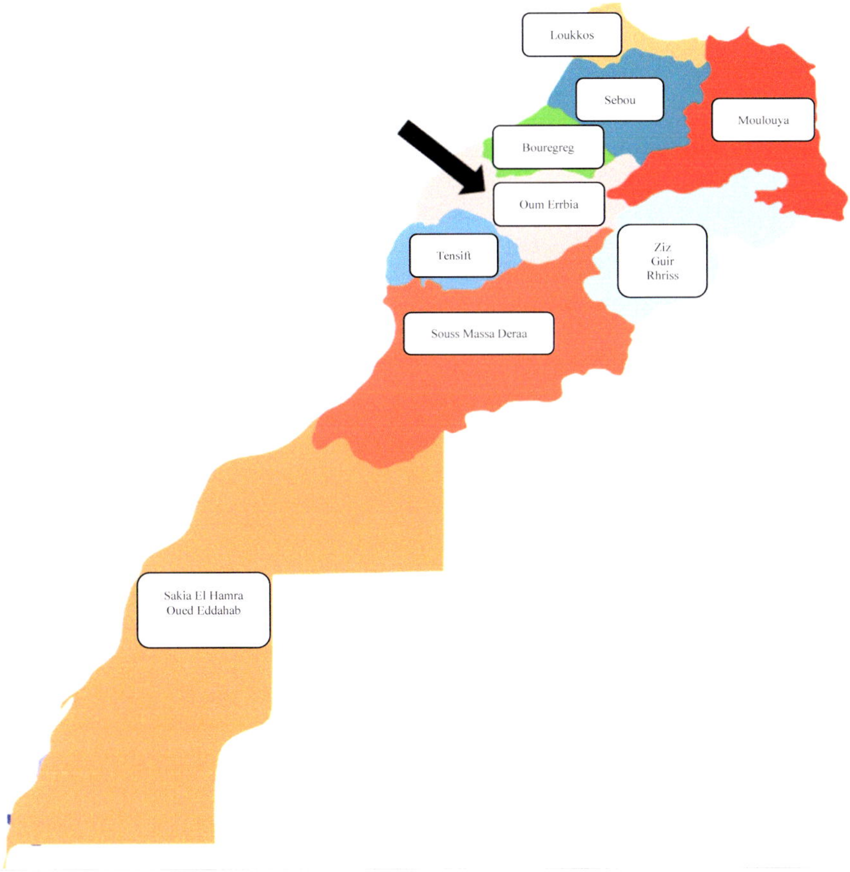

Fig. 6 Map of the geographical location of the Oum Errbia watershed (DGH 2023)

in the Middle Atlas Mountains, and up to an arid climate in the Rhamna plain and semi-arid in the Tadla plain (Arif and Doumani 2012).

Mainly focusing on agriculture and livestock, the basin boasts the country's largest irrigated area, encompassing 494,000 ha, with over 90% dedicated to cereal farming.

The major hydraulic infrastructures of the basin include 16 large dams, used as water storage tanks with a total capacity of 5,096.6 million m^3. Some of these dams also contribute to hydroelectric production, with a total capacity of 760 MW, thus representing 70% of the national hydroelectric park. These structures play a crucial role in storing water during periods of floods, regulating its evacuation to meet the needs of agriculture and irrigation, while providing water during periods of drought (Arif and Doumani 2012).

The maps produced by our GIS application highlight areas strongly affected by siltation in this watershed.

Due to the significant erosions observed in the drained basins and the substantial volumes of sediments transported by the rivers, these dams have experienced siltation phenomena, thus causing a reduction in their storage capacity (Arif and Doumani 2012). This partly explains the heavily silted areas highlighted by our GIS application.

On the other hand, various factors, such as the pronounced topography, the limited vegetation cover, and the sensitivity of the soils to erosion, influence soil erosion in the Oum Errbia watershed. Studies reveal that this watershed is highly subject to erosion, with soil losses ranging from 50 to 400 tons per hectare per year, thus affecting 54% of the surfaces. Only 9% of the catchment area has losses below the tolerance threshold (<7 tons per hectare per year) (Yjjou et al. 2014).

Erosion factors accentuate this deterioration and act in concert to accelerate the phenomenon. These factors include steep slopes on 62% of the catchment area, an alarming decline in vegetation cover, as well as the presence of highly erodible soils (59% of soils have a K factor > 0.4) (Yjjou et al. 2014).

4.2 Additional Contributions of the Developed GIS Application

In addition to the spatiotemporal analysis of dam siltation, the GIS application that we have developed makes significant additional contributions by including in its database for each dam, construction details, hydrological data, inspection results, auscultation data and maintenance operations.

These extended functionalities reinforce the usefulness of the application in the global management of dams and offer a complete and detailed vision of the critical aspects related to the construction, monitoring, and maintenance of hydraulic infrastructures.

The functionality of the GIS application is centered around the "dams" menu, illustrated in Fig. 7. This feature empowers users to seamlessly integrate information for new dams or make modifications to the existing 364 dams already cataloged

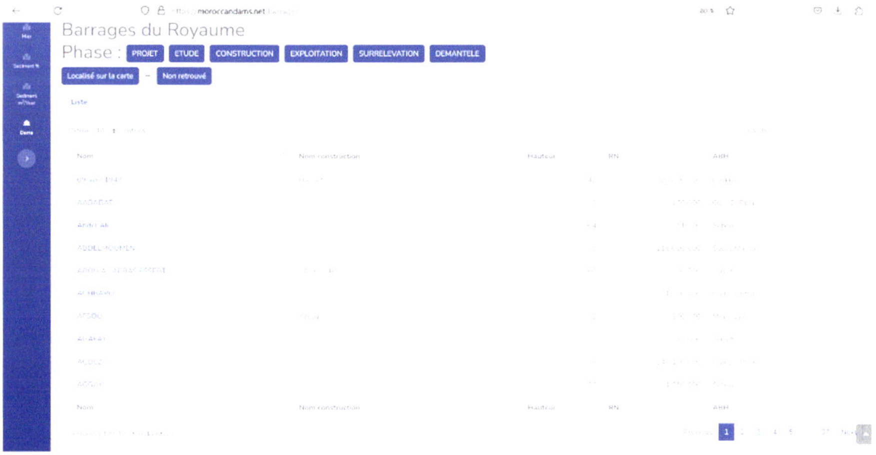

Fig. 7 "Dams" menu of the developed GIS application

within the system, spanning a range from large dams to smaller ones. The application's seamless integration with a purpose-built database ensures a comprehensive repository of detailed information pertaining to dams.

This purpose-built database encompasses a wealth of information, including specifics related to dam construction, hydrological parameters, inspection findings, and any other relevant data associated with the dams' operation. Users, whether researchers, policymakers, or dam managers, can efficiently manage and update this wealth of data through the user-friendly interface of the GIS application.

Navigating the application, users can explore the "map" tab, providing an exhaustive list that encompasses all dams cataloged in the database. This comprehensive list serves as a quick reference for users to access general information about each dam, fostering a holistic understanding of the dam landscape in Morocco.

The second tab within the application creates a cartographic representation, offering an insightful visualization of sediment accumulation rates for each dam. This visual representation aids in quickly identifying patterns and trends related to siltation across different dams, enabling users to make informed decisions about potential interventions or targeted monitoring efforts.

The third tab delves into cumulative sediment volumes within each reservoir. This provides a quantitative perspective on the extent of siltation over time, allowing users to gauge the overall impact on storage capacity and assess the urgency of potential remedial actions.

Essentially, the robust features of the GIS application under the "dams" menu not only facilitate data management but also empower users with valuable tools for analysis and decision-making related to dam reservoirs in Morocco.

4.3 Implications for Water Resource Management and Siltation Mitigation Strategies

The outcomes derived from our spatiotemporal analysis of dam reservoirs siltation in Morocco carry substantial implications for regional water resource management. The gradual accumulation of silt in reservoirs can profoundly impact dam storage capacities, water quality, flow regulation, and the overall durability of these critical infrastructures. In this section, we will delve into potential measures and strategies to counteract the impact of siltation and ensure the sustainable operation of dams.

1. **Proactive Siltation Management**: A proactive approach involves integrating siltation prevention measures into dam design and planning. This may encompass the establishment of upstream settling ponds to capture suspended sediments before they reach the primary reservoir. Encouraging protective vegetation and sustainable agricultural practices in watershed areas can also mitigate sediment input.
2. **Routine Maintenance and Dredging**: Regular reservoir maintenance, including the removal of accumulated sediments by dredging, is one of the most common siltation management strategies. This process necessitates careful planning to minimize environmental disruption while maintaining sufficient storage capacity. Dredged sediments can potentially be reused for land restoration projects or other beneficial applications (Mohafid et al. 2021, 2023).
3. **Controlled Erosion Techniques**: The implementation of controlled erosion techniques within watersheds can diminish rainwater runoff and the sediment load carried to dam reservoirs. Methods may involve establishing vegetated buffer zones, constructing upstream dams to slow water flow, and stabilizing slopes (Tisserant et al. 2021).
4. **Watershed Assessment**: A comprehensive comprehension of sediment sources within watersheds is crucial for effective siltation management. Detailed hydrological and geological investigations can pinpoint high-risk sediment intake areas, allowing for more precise targeting of management actions (Dai et al. 2004).
5. **Awareness and Education**: Engaging local stakeholders and neighboring communities in siltation management can promote responsible and sustainable practices. Promoting awareness and education on the consequences of siltation and preventive measures can encourage active participation in water resource preservation.

In summary, the spatiotemporal analysis of dam reservoirs siltation offers crucial insights for making informed decisions in water resource management. The imperative lies in adopting an integrated, multidisciplinary approach that engages government entities, hydrology experts, engineers, ecologists, and local communities. This collaborative effort is vital to effectively mitigate the effects of siltation and ensure the enduring sustainability of dams in Morocco.

4.4 Prospects for Future Research

The spatiotemporal examination of dam reservoirs siltation in Morocco has set the groundwork for future investigations, aiming to enhance our comprehension of this complex phenomenon and its implications for water resources and the environment. The findings prompt several inquiries and areas for exploration that merit closer examination.

1. **Predictive Siltation Modeling**: Proactive siltation management can benefit from predictive modeling based on bathymetric data. Subsequent studies may concentrate on creating numerical models capable of anticipating long-term siltation patterns, and integrating climatic, geological, and hydrological factors (Hasbaia et al. 2017).
2. **Ecological Consequences Studies**: The exploration of siltation's impacts on aquatic ecosystems and riparian habitats remains relatively unexplored. Future research efforts could delve into the ecological repercussions of siltation, encompassing effects on biodiversity, water quality, and food chains.
3. **Socio-Economic Impact Assessment**: A comprehensive understanding of the socio-economic impacts of siltation on local communities, agriculture, energy production, and other sectors is imperative. Economic and social studies could quantify the costs and benefits linked to siltation and de-siltation, as well as identify best practices to minimize disturbances.
4. **Integrated Management Studies**: Recognizing that siltation is closely linked to broader watershed dynamics, forthcoming research could adopt an integrated management approach. This involves examining siltation in the context of interactions between human activities, land use, runoff, and geological processes.
5. **Analysis of Innovative Management Solutions**: Given the continued evolution of technologies and management approaches, future studies might explore innovative solutions to mitigate siltation. This could involve investigating the use of more sustainable dredging materials, implementing sediment stabilization techniques, and promoting sustainable agricultural practices in watersheds (Mohafid et al. 2021, 2023).
6. **Regional Comparative Studies**: Comparative analyses of siltation patterns across different regions or countries offer insights into local factors and best management practices. Such studies can aid in identifying transferable approaches and understanding the influences of geological and climatic conditions.

Essentially, the spatiotemporal analysis of dam reservoirs siltation in Morocco opens avenues for complementary research. A comprehensive understanding of siltation and its impacts necessitates a multidisciplinary approach, involving researchers in hydrology, geology, ecology, economics, and engineering, as well as collaborations with dam managers and local communities.

5 Conclusion

The spatiotemporal analysis powered by the integration of bathymetric and GIS data, provided crucial information on the siltation dynamics within Moroccan dam reservoirs. By coupling historical bathymetric measurements with GIS technologies, we have meticulously mapped the spatiotemporal models of siltation, highlighting the various factors influencing this complex phenomenon.

Our results have highlighted significant siltation trends across various large dams in Morocco, focusing on distinct spatial and temporal variations. The exhaustive analyses allowed us to identify areas characterized by an increased accumulation of sediments and to highlight regions most vulnerable to this phenomenon.

Beyond contributing to the scientific understanding of siltation, our research has substantial implications for the sustainable and integrated management of water resources in Morocco. It provides information necessary to decide on mitigation measures aimed at preserving the storage capacity of dams and optimizing their use for water supply, irrigation, and energy production.

However, our study faces certain limitations, mainly related to the availability of historical data and the intermittent nature of accessible bathymetric measurements. These constraints underline the importance of continuous monitoring and continuous data collection efforts to achieve a more complete understanding of dam siltation in the future.

The spatiotemporal analysis of the siltation of dams in Morocco lays the foundations for future research. These activities aim to deepen the complexities of this phenomenon and to comprehend its implications for water resources and the environment. The results of our research have revealed many investigations and areas of research that merit closer examination.

References

Apiag C, Cadiz E, Lincopinis D (2023) A review on PHP programming language, Western Mindanao State University, Pagadian City, Philippines, ResearchGate, May 2023. https://www.researchgate.net/publication/371166635

Arif S, Doumani F (2012) Morocco, cost of the degradation of water resources in the Oum Er-Rbia Basin. Sustainable Integrated Water Management Program (SWIM-SM) funded by the European Commission, Brussels. https://swim-sm.eu/

BST (2023) The British Columbia Institute of Technology (BCT) company, owner of CodeIgniter, Official website. https://github.com/bcit-ci/CodeIgniter. Accessed 08 Dec 2023

Dai JJ, Lorenzato S, Rocke DM (2004) A knowledge-based model of watershed assessment for sediment. Environ Modell & Softw 19(4):423–433. https://doi.org/10.1016/S1364-8152(03)00158-0, ISSN 1364–8152

DGH (2023) General Directorate of Hydraulics/Moroccan Ministry of Equipment and Water, Hydraulics in figures, June 2023, p 12. https://www.equipement.gov.ma/eau/documentation/Pages/accueil.aspx

Hasbaia M., Zeroual S., Dougha M., Paquier A., Poulard C (2018) Prediction of dams silting in semi-arid region using erosion map under GIS environment, case of Ksob watershed in Hodna Region

(Algeria). In: Recent advances in environmental science from the Euro-Mediterranean and surrounding regions. EMCEI 2017. Advances in science, technology and innovation. Springer. https://doi.org/10.1007/978-3-319-70548-4_229

Lahlou A (1994) Siltation of dams in Morocco. Wallada Publications, Casablanca, 286p

Masih I, Maskey S, Mussá FEF, Trambauer P (2014) A review of droughts on the African continent: a geospatial and long-term perspective. Hydrol Earth Syst Sci 18:3635–3649. https://doi.org/10.5194/hess-18-3635-2014

Mohafid S, Stour L, Agoumi A, Ouassil M (2021) Guide for valuing mud extracted from dam reservoirs in Morocco. In: E3S web of conferences, vol 314, p 01001. https://doi.org/10.1051/e3sconf/202131401001

Mohafid S, Stour L, Agoumi A (2023) Optimization of the methodology for valorization of mud retained by dams: Morocco case study. Glob J Eng Technol Adv 17(01):055–068. https://doi.org/10.30574/gjeta.2023.17.1.0196

Moroccan Ministry of Equipment and Water Homepage (2023). https://www.equipement.gov.ma/Pages/accueil.aspx. Accessed 19 Oct 2023

OSGF (2023) The Official website of Open Source Geospatial Foundation owner of OpenLayers. https://www.osgeo.org/projects/openlayers/. Accessed 08 Dec 2023

Remini B, Hallouche W (2005) Prevention of siltation in Maghreb Dams. Larhyss J 04:69–80. ISSN 1112-3680. © 2005 Research Laboratory in Underground and Surface Hydraulics

Tisserant M, Bourgeois B, González E, Evette A, Poulin M (2021) Controlling erosion while fostering plant biodiversity: A comparison of riverbank stabilization techniques. Ecol Eng 172:106387. https://doi.org/10.1016/j.ecoleng.2021.106387, ISSN 0925-8574

Walling Des E (2012) The role of dams in the global sediment budget, erosion, and sediment yields in the changing environment. In: Proceedings of a symposium held at the Institute of Mountain Hazards and Environment, vol 356, CAS-Chengdu, China, 11–15 October 2012. IAHS Publications

Yjjou M, Bouabid R, El Hmaidi A, Essahlaoui A, El Abassi M (2014) Modeling of water erosion via GIS and the universal soil loss equation at the level of the watershed Oum Er-Rbia. Int J Eng Sci (IJES) 3(8):83–91 (2014). ISSN (e): 2319 – 1813, ISSN (p): 2319 – 1805

Yobom O (2020) Climate change, agriculture and food security in the Sahel. Economics and finance. Université Bourgogne Franche-Comté (2020)

Artificial Intelligence

Machine Learning Based Statistical Downscaling Approach for the Assessment of Climate Change Impact on Precipitation in Damaturu, Nigeria

Jazuli Abdullahi, Ala Tahsin, Mehmet Irfan Yesilnacar, Abdullah İzzeddin Karabulut, and Oluwatoyin Daramola

Abstract Precipitation forecasts for the future are crucial for the efficient management of water resources. These forecasts are often made using global circulation models (GCMs). Therefore, this paper examined the impact of climate change on precipitation for Damaturu, Yobe State, Nigeria between 2050 and 2080 using GCM variables. To achieve this, an artificial neural network (ANN) was utilized to downscale observed precipitation data using the BNU-ESM GCMs under the emission scenario RCP 4.5. The mutual information (MI) technique was used to rank various climate predictors based on their influence on precipitation. In order to downscale the precipitation data, five distinct predictor combinations were used to create the ANN models. Next, the Root Mean Square Error (RMSE) and Determination Coefficient (DC) performance metrics were used to identify the best downscaling model. It was discovered that M1, which combined the top 8 ranked predictors, performed the best throughout both the projection and downscaling phases. The final M1 results indicated that a decrease in precipitation will likely be experienced in the Damaturu region within the given period. In the months with the highest amounts of precipitation, the decrease will be more pronounced with the greatest amount of up to 20% occurring during the wettest month of August, near the end of the twenty-first century.

Keywords Climate change · Mutual information · Global circulation models · Damaturu region · Nigeria

J. Abdullahi (✉) · O. Daramola
Civil Engineering Department, Faculty of Engineering, Baze University, Abuja, Nigeria
e-mail: jazuli.abdullahi@bazeuniversity.edu.ng

A. Tahsin · M. I. Yesilnacar
Environmental Engineering Department, Harran University, Haliliye, Sanliurfa, Türkiye

A. İ. Karabulut
Remote Sensing and Geographic Information Systems, Harran University, Sanliurfa, Türkiye

N. Rebai et al. (eds.), *GIS, Applied Computing and Data Science for Water Management*, Lecture Notes in Geoinformation and Cartography,
https://doi.org/10.1007/978-3-031-63038-5_11

1 Introduction

The water resource's temporal and spatial variability has been greatly influenced by climate change, thus resulting in water resources assessment for catchments being highly important (Pascual et al. 2015). The General Circulation Models (GCMs) are thought to be the most sophisticated tools available for climate change forecasts of hydroclimatic variables at the global scale (Bates et al. 2010). GCMs are compelled to incorporate plausible future GHG emission scenarios for the development of global climate scenarios that are likely to occur in the future. Due to their function at a coarse spatial scale, the sub-grid size processes including land surface dynamics and cloud physics are unable to be resolved by the GCMs. However, Earth topography is coarsely represented within the structure of GCMs (Iorio et al. 2004). Consequently, projections of GCMs including modelling of water resources allocation and hydrologic modelling in catchment scale applications cannot be readily used.

The statistical as well as dynamic downscaling techniques have been developed to close the gap in spatial scale between the GCM outputs of coarse scale and local scale hydroclimatic variables (Wilby and Wigley 1997). The conventional relationships between hydroclimatic variables in the catchment scale and GCM outputs are statistically developed in statistical downscaling (Benestad et al. 2008). In dynamic downscaling, local or catchment scale variables and GCM outputs are bridged using physics-derived equations (Fowler and Wilby 2010). In contrast to dynamic downscaling, the statistical downscaling approach has become increasingly popular because to its ease of use and cheap computing cost (Sachindra et al. 2016).

Artificial neural network (ANN), a type of machine learning (ML) model, have been successfully used in recent decades to forecast complex hydro-climatological characteristics. For example, (Nourani et al. 2019a) used ANN and other methods to estimate Iraq's daily global solar radiation. (Tahsin et al. 2023) used a Gaussian Processing Regression model based on machine learning to do a spatiotemporal prediction of reference evapotranspiration in the Ara-ban Region of Turkey. Numerous machine learning (ML)-based researches are available in the literature that address statistical downscaling and future projections of climate variables. In order to downscale precipitation over the Australian State of Victoria, (Sachindra et al. 2018) used ANN and a number of machine learning techniques. (Elkiran et al. 2021) took into account ANN, Multiple Linear Regression (MLR), and Adaptive Neuro-fuzzy Inference System (ANFIS) for statistical downscaling and future temperature and precipitation projections throughout Cyprus's Nicosia, Kyrenia, and Famagusta stations. (Junaid et al. 2023) used the Statistical Downscaling Model (SDSM) and machine learning techniques to downscale the extremes of temperature and precipitation for Prince Edward Island (PEI). In Nigeria, there is very little use of machine learning (ML) models, despite their widespread acceptance and effective applications for downscaling climatic variables. Therefore, the purpose of this study was to determine whether the ANN model could be used to downscale and predict future changes in precipitation for the Nigerian station of Damaturu.

2 Materials and Methods

2.1 Study Area

Northeastern Nigeria's Yobe state, with its capital city of Damaturu, is known for its tropical climate, which features distinct dry and wet seasons. A significant wetland is often used to describe Yobe State in the semi-arid region. The tropical vegetation of the region was typified by several edulious plants, scattered shrubs, tamarind, and acacia species (Ambrose et al. 2018).

2.2 Study Data

The average monthly precipitation forecast and data for the years 1990–2020 were collected from the Meteorological Agency of Nigeria for this study. Furthermore, for the same 30-year timeframe and a grid point situated at Damaturu, 15 large-scale GCM predictors utilized under Representative Concentration Pathway (RCP) 4.5 for the BNU–ESM GCM, were retrieved from http://cera-www.dkrz.de. Table 1 displays the statistics of the precipitation data, Fig. 1 shows the map of Nigeria and Damaturu study area, and Fig. 2 shows the monthly precipitation over the research area.

As seen in Table 1, the monthly precipitation within the base period (1990–2020) can be as low as 0, indicating that there are periods with no rainfall. This especially happens during dry season when there is no precipitation for some months. However, the maximum cumulative precipitation in the wettest is 272.65 mm/month with an average value of 50.33 mm/month.

From the monthly time series of precipitation shown in Fig. 2, it can be deduced that the Damaturu region features a precipitation pattern that falls in line with that of a tropical Savannah climate, with wet seasons falling in warmer months and dry seasons falling in cooler months. The observed precipitation pattern for the region can also be attributed to the region experiencing the seasonal effects of monsoons, making it extremely wet during monsoon months and extremely dry during other months—with very few or no months displaying moderate amounts of precipitation. Due to the constant movement of water from the earth to the atmosphere through the concept of the water cycle, the precipitation observed in the Damaturu is comprised of water that evaporated from both within the region and adjacent regions. It can therefore be deduced that any significant variation in peak annual precipitation values of the region overtime, are influenced by climatic and/or geographical events/changes that

Table 1 Precipitation data descriptive statistics

Station	Parameter	Unit	Min	Max	Average	St. deviation
Damaturu	Precipitation	mm	0	272.65	50.33	70.5

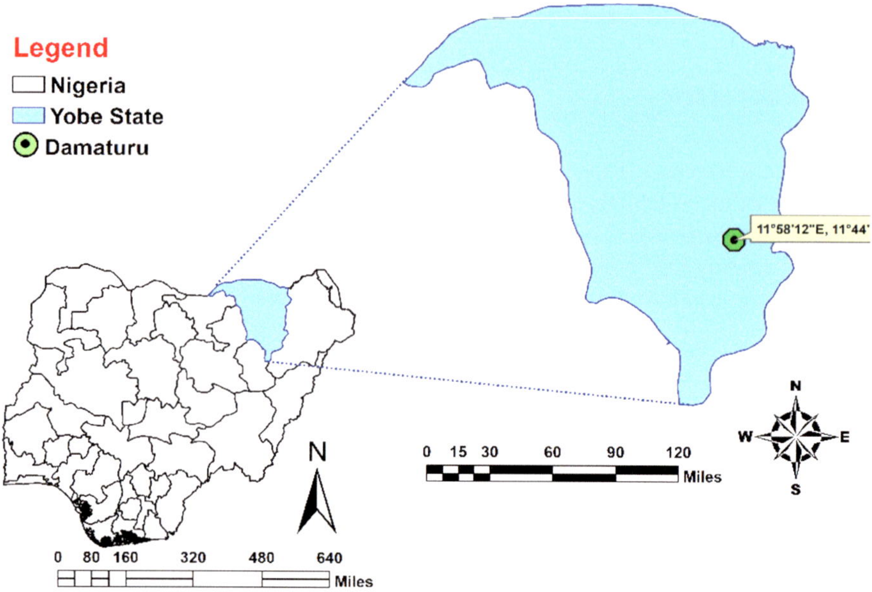

Fig. 1 Damaturu study area

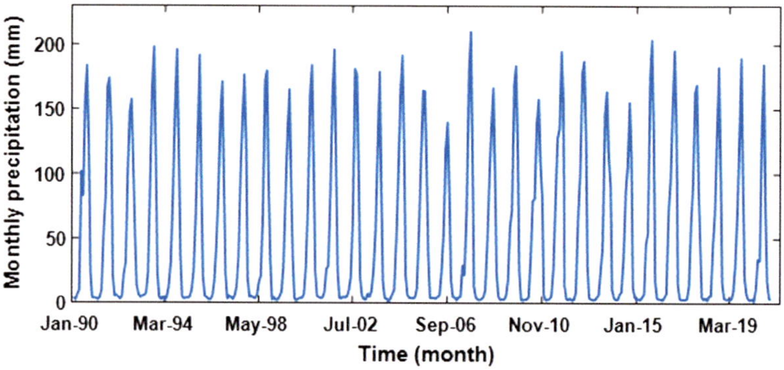

Fig. 2 Monthly precipitation distribution from 1990 to 2020 for the study station

occur both within and outside the region. The GCM data employed for the study is described in Table 2.

As Table 2 illustrates, the predictors utilized in this study are connected to parameters influencing precipitation, such as temperature, humidity, wind speed, and other factors at both surface and near surface heights.

Table 2 Data description of large scale GCM predictors

S/N	Predictors	Description	Height
1	hurs	Relative humidity	Near surface
2	hfls	Surface_upward_latent_heat_flux	Surface
3	huss	Specific humidity	Near surface
4	uas	Zonal wind speed	Near surface
5	prw	Atmospheric water vapor content	Surface
6	vas	Meridional wind speed	Near surface
7	evspsblveg	Transpiration	Surface
8	prc	Precipitation	Surface
9	evspsbl	Water evaporation	Surface
10	mrros	Surface_runoff_amount	Surface
11	tasmin	Minimum air temperature	Surface
12	mrso	Atmosphere_water_content	Near Surface
13	croot	Root_Carbon_Content	Surface
14	ts	Sea surface temperature	Surface
15	sfcwind	Wind speed	Near surface

2.3 Proposed Methodology

ANN was used in this study to do the downscaling modeling and future precipitation projection. As a result, the proposed method was developed in three stages.

2.4 Performance Indices of the Models

The root mean square error (RMSE) and determination coefficient (DC) performance indices were employed to assess the efficiency and accuracy of the models. They are provided as follows:

$$DC = 1 - \frac{\sum_{i=1}^{N}(p_i - \widehat{p_i})^2}{\sum_{i=1}^{N}(p_i - \overline{\overline{p_i}})^2} \quad (1)$$

$$RMSE = \sqrt{\frac{\sum_{i=1}^{N}(p_i - \widehat{p_i})^2}{N}} \quad (2)$$

where N, p_i, $\overline{\overline{p_i}}$ and $\widehat{p_i}$, are the number of observations, observed precipitation ith value, mean of the observed values and predicted ith value, respectively. The DC has a value between negative infinity and 1, and as it approaches 1, the model's efficiency increases. As the RMSE number approaches zero, it increases in model efficiency and spans from 0 to infinity (Nourani et al. 2019b).

2.5 *Artificial Neural Network (ANN)*

A machine learning model called an artificial neural network (ANN) is based on biological neurons seen in human anatomy (Ibrahim and Tulay 2023). You may find more information on the use of ANN in the study (Ibrahim and Tulay 2023).

3 Results and Discussions

Three (3) phases were employed in this study. In phase 1, sensitivity analysis was carried out and the most appropriate predictors for producing the best downscaling models were determined using Mutual Information. ANN downscaling method was applied in phase 2 of the study to statistically downscale the precipitation data. Finally, in phase 3 of the study, the projection of future precipitation for the study region was conducted. The results of the study are presented in the following section.

Phase 1 results: Sensitivity analysis

Due to the greenhouse effect, the selection of the dominant predictors is a necessity when performing downscaling modeling within fields related to climate change such as precipitation. According to Nourani et al. (2019a), to obtain the best performance from a data driven model, the selection of dominant predictors should follow assignment of the GCM. Therefore, for the study station, the MI feature extraction technique was applied to a total of 15 atmospheric predictors generated from the BNU-ESM GCM at Damaturu station. The results of the applied predictor screening are shown by Fig. 3.

Based on the MI technique results displayed above, a larger percentage indicates better performance for the variables.

As shown in Fig. 3, the higher the MI of a predictor, the more dominant it is in relation to the predictand (precipitation). Therefore, for the Damaturu station, the results show that prw (8%), tasmin (8%), and prc (8%) are respectively most dominant predictors in relation to precipitation, each having a strong nonlinear relationship with precipitation. These predictors are related to atmospheric moisture and temperature. This can be due to the semi-arid climate of the region, as during warmer months (i.e. higher atmospheric temperature) which experience the most precipitation, the air tends to hold more moisture (i.e. higher atmospheric water vapor content), and during the colder months (i.e. lower atmospheric temperature) which experience the least precipitation, the air tends to hold less moisture (i.e. lower atmospheric water vapor content). It should also be noted that these predictors along with precipitation

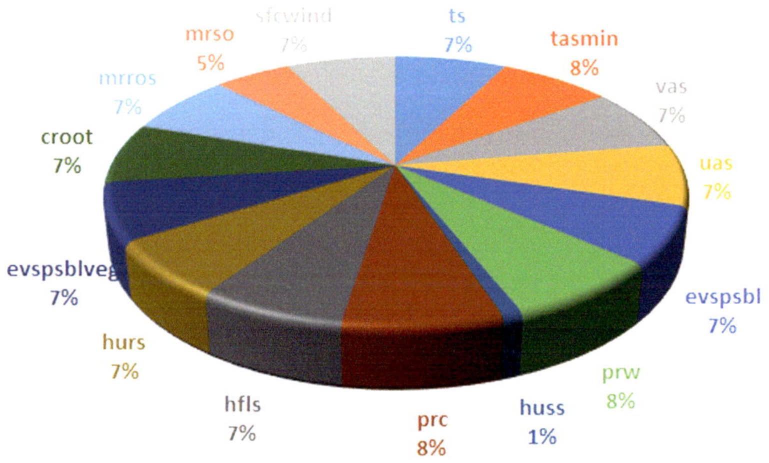

Fig. 3 Mutual information (MI) results for the predictor variables

are important parameters of the hydrologic water cycle. Next, statistical downscaling modeling was carried out after determining the dominant predictors.

Phase 2 results: Results of the ANN downscaling

The results of the statistical downscaling modeling of precipitation performed using the ANN model are presented in this section. In this study, several ANN models were developed using a combination of the top 3, 4, 5, 6, and 8 ranked predictors (see Table 3), respectively, as inputs. The validation method applied to each model was a basic train/test split validation. Following this method, the data for the relevant predictors in each model were randomly split into two subsets, with one subset containing 75% of the data for training the model and the other containing 25% of the data for testing/validating the model. Several model algorithms were also considered to produce the best performing model, and a trial & error approach was adopted to select the number of hidden layer neurons, which are very important in AI based downscaling modeling. Table 3 shows the results of the ANN downscaling.

Table 3 displays the evaluation of each model's downscaling performance based on the DC and RMSE performance metrics. The "goodness of fit" (or DC) of the anticipated precipitation to the actual or observed precipitation for each model configuration is indicated by the RMSE, which represents prediction errors. According to the validation results, the most effective downscaling model for both performance metrics was determined to be the ANN model M1, which employed the top 8 rated predictors as inputs. With respect to DC, model M1 has 0.26% more efficiency than the next most efficient model M2, and 4% more efficiency than the least efficient model M4, while with respect to RMSE, model M1 can be seen to have the least prediction error. It should also be noted that the RMSE values for the training and validating subsets of model M1 are very similar, suggesting the model will retain its predictive value when tested out of sample.

Table 3 Results of the ANN downscaling

Model	Inputs	Structure	Training		Validation	
			DC	RMSE	DC	RMSE
M1	ts, uas, prw, sfcwind, mrros, tasmin, prc, vas	8-17-1	0.8310	0.1070	0.8266	0.1044
M2	Prw, vas, ts, uas, tasmin, prc,	6-15-1	0.8274	0.1081	0.8240	0.1052
M3	prc, vas, ts, prw, tasmin	5-10-1	0.8342	0.1060	0.8239	0.1052
M4	tasmin, prc, Prw, vas	4-8-1	0.8302	0.1072	0.7863	0.1159
M5	prc, Prw, tasmin	3-9-1	0.8162	0.1075	0.8050	0.1149

Overall, it was found that improved precipitation downscaling efficiency was achieved by gradually increasing the number of ranking predictors used as inputs in the ANN model. However, exceptions to this observation occurred anytime a wind related predictor was newly added to the inputs of the model. This is displayed in the results (Table 3) where model M5 outperformed M4, which used the same predictors as M5 with an addition of meridional wind speed (vas) and where M2 and M3 showed virtually no variation in performance despite the inclusion of zonal wind speed (uas) to M2, suggesting that wind speed is not a strong predictor of precipitation within the study region. This could be attributed to the climate of the region, as the prevailing wind (harmattan wind) brings little to no moisture and is most active in the region during the cooler months where precipitation is at its lowest.

The time series from 1990 to 2020 of the monthly observed precipitation and the downscaled time series of the precipitation within the mentioned period by each model were compared to assess the monthly ANN downscaling performance of the precipitation as shown in Fig. 4.

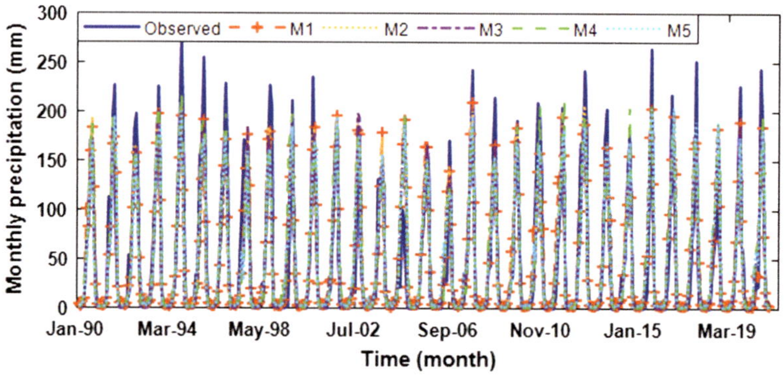

Fig. 4 Comparison of the observed and downscaled monthly precipitation time series

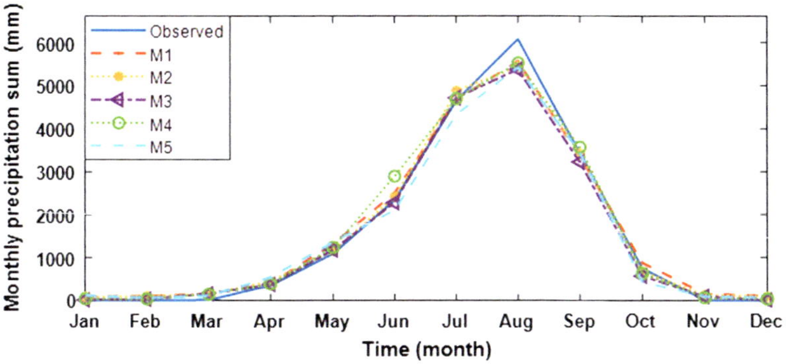

Fig. 5 Performances comparison of the developed models

Figure 4 illustrates the efficiency of the 5 different model combinations for the statistical downscaling of precipitation. With values fitting (close) to the observed values, higher performance is achieved. However, similar to the tabular results presented (Table 3), M1 performed better than M2–M5, with values more fitted to the observed values. From both Table 3 and Fig. 4 shown, it can be deduced that for due to the uncertain and stochastic nature of precipitation more variables are needed to unveil its dynamism and provide good downscaling capabilities by ANN.

Figure 5 demonstrated the monthly downscaling capabilities of the developed models.

It is evident from Fig. 5 that each model generated a downscaled monthly precipitation time series pattern that was comparable to the observed monthly precipitation time series; however, each model generally underestimated the observed peak precipitation values that occurred throughout the time series (that is, the observed pattern's peak values for each parabolic formation were, on average, higher than the downscaling models' patterns). For the research station, the months with the least amount of precipitation are January and December, whereas the months with the maximum precipitation are June through September, as Fig. 5 illustrates. Models are more dependable during low precipitation times, based on how well the downscaled models fit and resemble the observed precipitation between January and December. It makes sense to think this way because more precipitation would make it harder to track and record the linear and non-linear processes that characterize the precipitation phenomenon. This will lead to less effective downscaling of precipitation with more fluctuations.

Phase 3 results: Future projection precipitation

This section presents the 30-year (2050–2080) precipitation estimate for the future. The anticipated precipitation in the future, based on the models that were used and the percentage change in precipitation, is shown in Figs. 6 and 7.

Figures 6 and 7 show the changes and percentage of changes respectively, in the monthly precipitation sum over the study time periods (i.e.1990–2020 for base period

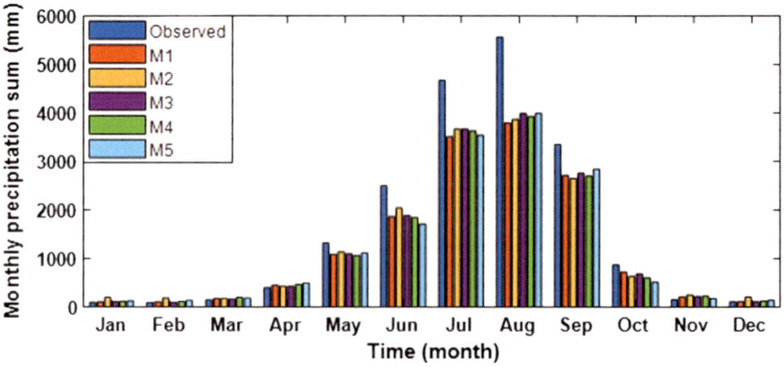

Fig. 6 Past and future projected precipitation

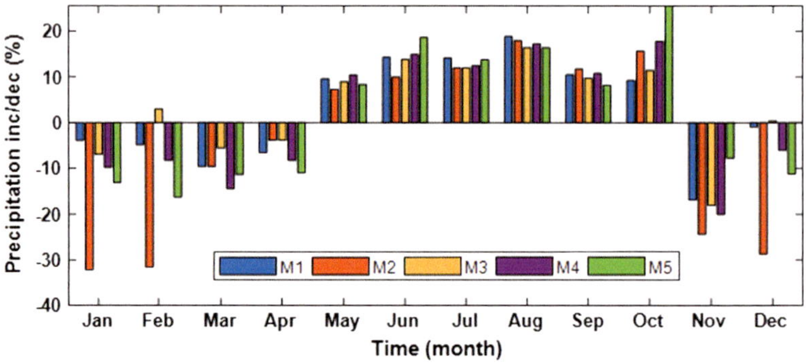

Fig. 7 The percentage change in precipitation between 2050 and 1980

and 2050–2080 for future period) based on the future projections of each model. It should be noted that positive values in the figures represent a decrease in precipitation while the negative values represent an increase in precipitation.

Figures 6 and 7 show that, for all models, precipitation decreased during the wetter months (May through October) and increased during the drier months. The only model that deviated from this pattern was model M3, whereby during the dry month of February, it showed a decrease in precipitation. The most decrease in monthly precipitation sum was seen in model M5 with a nearly 25% decrease during the month of October while the most increase was found in model M2, with an upward of 30% increase during the month of January.

According to the precipitation results for the future projection (Fig. 6), the model M1 displayed slightly better performance than the other models, therefore, discussions presented in this section are centered mainly around the comparison made between the future projection by the model M1 and benchmark precipitation. In line with the previously mentioned general pattern, model M1 also showed an increase in precipitation during drier months and a decrease during wetter months. The largest decrease, of 20%, happened in August, while the largest increase, of roughly 17%, happened in November.

4 Conclusion

This article examined the future changes in monthly precipitation that may likely occur owing to global climate change using ANN downscaling for Damaturu station Yobe, Nigeria over the years 2050–2080. The results revealed that Damaturu region will experience an increase in monthly precipitation during drier months and a decrease during wetter months according to the final M1 projections. The largest decrease, of 20%, is expected to occur in August, while the largest increase, of roughly 17%, is expected to occur in November. The difference between the precipitation time series produced by model M1 for the designated past and future time periods demonstrated that, in general, climate change will cause a decrease in precipitation in the Damaturu region, with this decrease being more pronounced in the months that receive the greatest amount of precipitation.

Competing Interests There is no competing interest among the authors.

References

Ambrose ZA, Umar Y, Binbol NL (2018) Glob J Pure Appl Sci 24:117–124
Bates BC, Walker K, Beare S, Page S (2010) Waterlines Rep Ser 28:196–198
Benestad, Rasmus E, Deliang C, Inger HB (2008) WSPC
Elkiran G, Nourani V, Elvis O, Abdullahi J (2021) J Hydroinformatics 23:1395–1415
Fowler HJ, Wilby RL (2010) Water Resour Res 46
Ibrahim Z, Tulay P (2023) J Abdullahi Environ Sci Pollut Res 30:3621–3643
Iorio JP, Duffy PB, Govindasamy B, Thompson SL, Khairoutdinov M, Randall D (2004) Clim Dyn 23:243–258
Junaid M, Afzaal H, Farooque AA, Abbas F, Wang X, Esau T (2023) Theor Appl Climatol 1–15
Nourani V, Elkiran G, Abdullahi J, Tahsin A (2019a) Nat Resour Res 28:1217–1238
Nourani V, Elkiran G, Abdullahi J (2019b) J Hydrol 577:123958
Pascual D, Pla E, Lopez-Bustins JA, Retana J, Terradas J (2015) Hydrol Sci J 60:2132–2147
Sachindra DA, Ng AWM, Muthukumaran S, Perera BJC (2016) Q J Roy Meteorol Soc 142:172–186
Sachindra DA, Ahmed K, Rashid M, Shahid S, Perera BJC (2018) Atmos Res 212:240–258
Tahsin A, Abdullahi J, Karabulut Aİ, Yesilnacar Mİ (2023) Adv Remote Sens 3:27–37
Wilby RL, Wigley TM (1997) Prog Phys Geogr 21:530–548

Towards Improved Rainfall Forecast Within the Ziz Basin Area: A Focused Exploration of Machine Learning Application

Sara Bouziane, Badraddine Aghoutane, Aniss Moumen, Ali Essahlaoui, Mohamed Hilali, and Anas El Ouali

Abstract Communities and industries rely heavily on accurate precipitation forecasts, which are essential for planning and decision-making across various sectors. Conventional methodologies for predicting weather entail a comprehensive understanding of physical phenomena, historical climate patterns, and the application of statistical models. Nevertheless, the intricate and nonlinear rainfall nature poses significant challenges to achieving precise predictions. In recent years, Machine Learning (ML) techniques have appeared as advantageous methods for enhancing the accuracy of rainfall forecasts. This article offers an extensive and comparative examination of classic techniques and evolved ML models relevant to precipitation estimation. By identifying the strengths and limitations of each approach, the study sheds light on the progress made in rainfall prediction using these innovative techniques. In addition, the study emphasizes the importance of pre-processing in improving predictive capabilities, exploring careful steps such as data cleaning, handling missing values, and scaling. Leveraging a comprehensive dataset collected from the Ziz Basin region, incorporating historical weather data spanning numerous

S. Bouziane (✉) · B. Aghoutane
Informatics and Applications Laboratory, Science Faculty of Meknes, Moulay Ismail University, Meknes, Morocco
e-mail: saray.bouziane@gmail.com; sa.bouziane@edu.umi.ac.ma

A. Moumen
Ibn Tofail University, National School of Applied Sciences, Kenitra, Morocco

A. Essahlaoui · M. Hilali
Laboratory of Geo-Engineering and Environment, Faculty of Sciences, Moulay Ismail University, Meknes, Morocco

A. El Ouali
Department of Environment, Functional Ecology and Environmental Engineering Laboratory, Faculty of Sciences and Technology, Sidi Mohamed Ben Abdellah University, Fez, Morocco

N. Rebai et al. (eds.), *GIS, Applied Computing and Data Science for Water Management*, Lecture Notes in Geoinformation and Cartography,
https://doi.org/10.1007/978-3-031-63038-5_12

climatic variables (precipitation, humidity, wind, temperature, and evaporation), our objective is to conduct a rigorous comparative analysis to assess the effectiveness of various ML models in enhancing rainfall estimation accuracy within this specific geographic area.

1 Introduction

Precisely predicting rainfall is critical to numerous facets of our daily existence, spanning agriculture, hydrology, urban development, transportation, and disaster preparedness (Bouziane et al. 2024). The socioeconomic ramifications of water resources are substantial (Barrera-Animas et al. 2022; Bouziane et al. 2024), covering a broad spectrum of effects, including infrastructure damage caused by floods, transportation system disruptions, challenges in water distribution, reservoir management issues, complications in irrigation scheduling, and ensuring equitable access to water resources (Bouziane et al. 2024).

Anticipated climatic changes are expected to intensify the frequency and severity of floods and other extreme weather occurrences in the coming years (Yucel et al. 2015; Bouziane et al. 2024). In alignment with this, our research builds upon the foundational insights presented in 'The Use of Big Data in Water Resources Management' (Bouziane et al. 2022) and 'Proposal of a Big Data System for an Intelligent Management of Water Resources' (Bouziane et al. 2021). These studies highlight the significant impact of advanced technologies, including Big Data, IoT, and Cloud Computing, in addressing water scarcity and augmenting accurate water resources governance. In this investigation, we narrow our focus to examine the implementation of Machine Learning (ML) technologies to enhance rainfall forecasting accuracy and foster more effective practices in water resource management.

The area of Ziz basin, renowned for its unique geographical and climatic attributes, stands as a cornerstone in influencing the socio-economic dynamics of its region (Bouziane et al. 2024). As we mentioned in our previous work, 'Enhancing Precipitation Prediction in the Ziz Basin: A Comprehensive Review of Traditional and Machine Learning Approaches.' (Bouziane et al. 2024), accurate rainfall prediction in this area holds indispensable significance for the welfare of local residents, endurable agriculture, efficient water resources management, and resilient disaster preparedness measures (Bouziane et al. 2024). Additionally, the abundance of historical weather data within this area positions it as a relevant and insightful case study, offering a valuable opportunity to investigate the potential of Machine Learning approaches in advancing the precision of weather forecasting (Bouziane et al. 2024). This article is an extended version of this paper, delving deeper into methodologies, preprocessing techniques, and insights derived from the unique geographical and climatic characteristics of the Ziz basin.

This research aspires to embark on a meticulous exploration of various methodologies in precipitation forecasting. Examining these practices, we aim to establish the basis for future endeavors (Bouziane et al. 2024). Mainly, we plan to leverage the meticulously collected and processed data to construct a strong and precise model for estimating precipitation in the specified site (Bouziane et al. 2024). This effort shows significant potential for increasing our comprehension of meteorological processes and enhancing the accuracy of our estimation (Bouziane et al. 2024).

The motivation of this study is the recognition that improved precipitation prediction contributes to scientific knowledge and carries practical implications across various sectors (Bouziane et al. 2024). We aim to facilitate forceful decision-making and drive improvement in meteorology (Bouziane et al. 2024).

The structure of this document is clear and comprehensively covers different aspects related to precipitation forecasting. It starts by discussing the impact of accurate precipitation forecasting on various sectors and identifying persistent challenges in predicting precipitation. Afterward, it thoroughly examines conventional methods employed in rainfall prediction and provides an overview of related works showcasing instances of various Machine Learning techniques applicable in this field. The emphasis is on models such as Artificial Neural Networks (ANN), Support Vector Machines (SVM), and Long-Short Term Memory Networks (LSTM) (Bouziane et al. 2024). The exploration further extends into the chronological and geographical setting of the study zone, providing valuable insights into the contextual elements influencing meteorological practices (Bouziane et al. 2024). Furthermore, this research outlines the precise methods used for data gathering and preparing, underscoring measures implemented to uphold the accuracy and reliability of the data (Bouziane et al. 2024). Each part contributes to the overarching objective of enhancing precipitation prediction accuracy.

2 Impact of Accurate Precipitation Forecasting on Different Sectors

Accurate precipitation forecasting holds profound implications for diverse sectors crucial to societal well-being and economic stability.

In agriculture, precise knowledge of impending rainfall allows farmers to optimize irrigation programs and strategically plan farming and gathering activities, contributing to increased crop yields and resource efficiency. Water resource management greatly benefits from accurate precipitation prediction, as it aids in reservoir planning and the efficient distribution of water supplies. Furthermore, industries reliant on weather conditions, such as construction and energy, can enhance operational planning and reduce risks associated with adverse weather events. Accurate forecasting also plays a pivotal role in disaster preparedness and response, enabling timely evacuation measures and resource allocation during extreme weather events. The healthcare sector, too, benefits from precise weather predictions by anticipating and preparing for weather-related health issues.

As such, the far-reaching impact of improved precipitation forecasting extends beyond meteorology, influencing strategic decision-making across various sectors, fostering resilience, and ultimately contributing to sustainable development.

Building on this and the crucial implications of accurate precipitation forecasting across sectors, it is imperative to recognize the persistent challenges associated with precisely predicting rainfall.

3 Challenges in Predicting Precipitation

Despite the transformative potential of accurate precipitation forecasting, numerous challenges persist in unravelling the intricacy and non-linearity of rainfall, shaping the landscape of meteorological research and forecasting methodologies.

Navigating the complexities of precipitation prediction presents challenges that continue to intrigue meteorologists and researchers alike. The non-linear dynamics of rainfall, influenced by myriad factors, pose a substantial hurdle for traditional forecasting methods. The variability in spatial and temporal scales further complicates the accurate precipitation events modelling. Additionally, uncertainties stemming from climate pattern variations and the increasing frequency of severe weather events, attributed to climate change, contribute to the intricacy of prediction models. Adequately capturing the microphysical processes within clouds and their interactions with the broader atmospheric conditions remains a persistent challenge. Moreover, the scarcity and irregular distribution of ground-based observation stations, especially in remote or less economically developed regions, limit the availability of comprehensive and high-resolution data.

As we delve into these challenges, it becomes evident that addressing these intricacies is vital for advancing the field of precipitation prediction, fostering resilience against the impacts of changing weather patterns and ensuring the reliability of forecasting systems in an era of dynamic climate shifts.

In light of this, a closer examination of traditional approaches to rainfall prediction reveals the evolving methodologies and insights that have shaped meteorological practices over the years.

4 Conventional Methods for Rainfall Forecast

In the past hundred years, weather prediction has been inconsistent, unpredictable, and untrustworthy (Bouziane et al. 2024; Lynch 2007). Rainfall forecasting typically depends on the combination of functioning numerical weather prototypes and exhaustive datasets from meteorological radars (Bouziane et al. 2024; Pham et al. 2020; Heistermann et al. 2013). These techniques analyze historical measurements to discern correlations between various meteorological factors, such as precipitation, geographic coordinates, pressure, temperature, relative humidity, and wind speed.) (Barrera-Animas et al. 2022; Bouziane et al. 2024; Pham et al. 2020).

Multiple studies utilized a variety of conventional and sophisticated methodologies, including regression analyses, climatology averages, numerical methods, and mathematical formulations (Bouziane et al. 2024; Ojo et al. 2022; Pham et al. 2020). For illustration, an empirical equation by AlHassoun is presented to estimate the intensity of rainfall in Riyadh, emphasizing the need for periodic reassessment of rainfall studies for maximum reliability (Bouziane et al. 2024; AlHassoun 2011). Similarly, to predict short-duration precipitation amounts within the Klang Valley region of Malaysia, Al Mamun et al. embarked on a study where they used statistical analyses and logarithmic graph fitting techniques to uncover potential correlations (Bouziane et al. 2024; Al Mamun 2018). This approach allowed them to capture the intricate relationships between various meteorological factors and rainfall patterns, offering insights into the localized dynamics of precipitation events. Furthermore, Villarini et al. delved into spatial and temporal modeling techniques as part of their investigation to mitigate uncertainties inherent in radar-based rainfall predictions (Bouziane et al. 2024; Villarini et al. 2014). By employing sophisticated mathematical and statistical models, they aimed to refine the accuracy of rainfall forecasts, particularly in regions where radar data may be prone to errors or inconsistencies due to geographical or atmospheric factors.

Nevertheless, the intricate and non-linear nature of rainfall presents significant challenges for accurate prediction (Barrera-Animas et al. 2022; Bouziane et al. 2024). To address these complexities, researchers have explored various methods such as Singular Spectrum Analysis (Bouziane et al. 2024; Niu et al. 2016), Empirical Mode Decomposition (Bouziane et al. 2024; Xiang et al. 2018), Wavelet Analysis (Barrera-Animas et al. 2022; Bouziane et al. 2024), and similar approaches. These methodologies aim to disentangle the intricate patterns and relationships within meteorological data to improve the accuracy of rainfall forecasts. However, these techniques often require substantial computational resources and prove time-consuming, ultimately leading to modest gains in prediction accuracy (Barrera-Animas et al. 2022; Bouziane et al. 2024).

In light of these limitations, Machine Learning approaches emerge as promising options. By leveraging algorithms capable of learning from data patterns and making predictions without being explicitly programmed, Machine Learning techniques hold the potential to overcome computational constraints and significantly enhance the precision of precipitation forecasts. These approaches enable researchers to extract meaningful insights from large datasets, uncover hidden patterns, and develop more robust models for rainfall prediction, thus paving the way for more accurate and reliable forecasts in the future.

5 The Potential of Machine Learning in Rainfall Prediction

In contemporary research endeavors, there has been a notable shift towards harnessing the potential of Machine Learning (ML) techniques to serve as predictive models for precipitation patterns (Barrera-Animas et al. 2022; Bouziane et al. 2024). This paradigmatic shift reflects a recognition of the limitations inherent in traditional forecasting methods and a growing appreciation for flexibility and predictive ability in ML algorithms. In our case study, various analyses and methodologies have paved the way for advancements in predictive modeling and data-driven approaches. This section explores related works and aims to provide a comprehensive overview of diverse ML applications, models, and techniques employed in rainfall prediction.

Artificial Neural Networks (ANN), a prominent Machine Learning technique, has demonstrated its efficacy in capturing intricate patterns and relationships within meteorological data (Narayanakumar and Raja 2016; Barrera-Animas et al. 2022). For instance, in a study by Pham et al. (2020), they developed an ANN model to predict rainfall patterns in Hoa Binh region, Vietnam. The model was trained on a comprehensive dataset, encompassing historical meteorological parameters such as daily rainfall, maximum temperature, minimum temperature, wind speed, relative humidity, and solar radiation (Pham et al. 2020). The trained ANN exhibited a remarkable ability to discern complex nonlinearities and dependencies, resulting in appropriate predictions of daily rainfall (Pham et al. 2020).

Continuing the exploration of Machine Learning techniques in rainfall prediction, Support Vector Machines (SVM) have emerged as a noteworthy approach (Evgeniou and Pontil 2001). In the same study conducted by Pham et al. (2020), SVM, along with other Machine Learning models, was employed to forecast daily precipitation patterns in Hoa Binh province, Vietnam. The study's outcomes emphasized the exceptional performance of SVM, outshining other artificial intelligence models in providing accurate and robust predictions for daily precipitation. Notably, SVM demonstrated effectiveness in addressing input variability by utilizing the Monte Carlo approach, further solidifying its role as a reliable and efficient tool for rainfall prediction (Pham et al. 2020).

Furthermore, Long Short-Term Memory (LSTM) networks have also demonstrated their effectiveness in weather prediction (Barrera-Animas et al. 2022), as highlighted in the study by Chao et al. (2018), Bouziane et al. (2024). This research highlighted the efficacy of LSTM networks in analyzing seasonal weather patterns and managing detector time series data, which are often prone to data disruptions (Bouziane et al. 2024; Chao et al. 2018).

As mentioned in the short version of this review (Bouziane et al. 2024), we contribute to a deeper understanding of the strengths and limitations of various approaches in improving the precision and efficiency of rainfall forecasts. Artificial Neural Networks (ANN), inspired by the structure and functionality of the human brain, excel at detecting intricate connections within rainfall data (Barrera-Animas et al. 2022; Bouziane et al. 2024; Pham et al. 2020). Support Vector Machines (SVM) specialize in organizing and interpreting paths in weather data, thereby enhancing the robustness of meteorological predicting models (Bouziane et al. 2024; Pham et al. 2020). LSTM networks, a subset of Recurrent Neural Networks (RNN), excel in catching long-term dependencies within periodical data, thereby proving invaluable for tasks related to rainfall forecasting (Bouziane et al. 2024; Chao et al. 2018).

As Machine Learning techniques in weather forecasting continue to advance, they play a crucial role in refining predictive models, ultimately leading to more accurate and reliable forecasts. These advancements are particularly significant for sectors such as agriculture, where precise weather predictions can optimize planting, irrigation, and harvesting schedules, thereby enhancing crop yields and resource efficiency. Additionally, in water resource management, accurate precipitation is paramount for ensuring the sustainable allocation and distribution of water resources, especially in regions prone to water scarcity. Furthermore, in disaster response, such as flood preparedness and mitigation, accurate weather forecasts enable authorities to implement timely and effective measures, minimizing potential damages and ensuring the safety of communities. Overall, the continuous development of Machine Learning techniques in weather forecasting holds immense promise for improving decision-making processes and enhancing resilience across various sectors reliant on weather-related information.

Transitioning to explore the study area and dataset, we describe the geographical context and meticulous data collection and pre-processing for robust meteorological analysis.

6 Study Area and Dataset

As we embark on a comprehensive examination of precipitation patterns and forecasting methodologies, this section delves into the intricacies of the study area and the dataset that form the core of our research. In advancing our understanding of rainfall dynamics, we begin by focusing on the specific geographical context of the Ziz Basin area. This distinctive region offers a microcosm of challenges and opportunities, providing valuable insights into the complexities of precipitation prediction. Afterward, we move on to the description of the dataset and the meticulous pre-processing operations taken to refine and prepare the data for our analytical pursuits.

6.1 Exploring the Geographical Significance of the Ziz Basin

In the vast expanse of the Central High Atlas, the Upper Ziz basin stands as a pivotal geographical entity, its geological and topographical features weaving a tapestry of significance within the region's broader landscape (Bouziane et al. 2024). Spanning an extensive area of approximately 14,411 square kilometers, the basin is defined by distinct geographical landmarks, including imposing massifs to the north and hills to the south (Bouziane et al. 2024; Mehdaoui et al. 2018) (Fig. 1). It holds particular

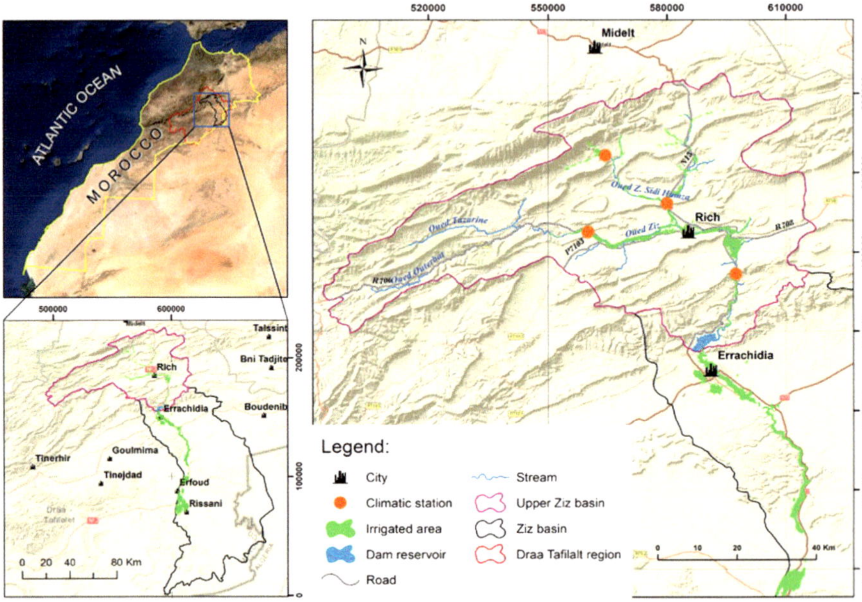

Fig. 1 Geological and topographical features of the Ziz basin (Bouziane et al. 2024)

importance within the broader High Atlas chain (Bouziane et al. 2024; El Ouali et al. 2023).

The geological and topographical characteristics of the Ziz watershed are not only of scientific interest; they underpin a complex web of social and cultural significance within the region (Bouziane et al. 2024). Communities living in this area rely on the resources provided by the basin, particularly the Ziz River, which supports agriculture, sustains ecosystems, and preserves cultural traditions. The basin's diverse landscapes and fertile lands along the riverbanks contribute to its rich cultural heritage (Bouziane et al. 2024).

Our study concentrates on the upstream Ziz Valley, a critical drainage basin covering approximately 4,351 square kilometers km^2 (Bouziane et al. 2024; El Ouali et al. 2023).

6.2 A Comprehensive Description of the Collected Dataset

In our data collection efforts within the study area, we encountered a noteworthy challenge stemming from the diversity in both temporal periods and available instrumentation across monitoring stations. Within the Ziz Basin, we identified four monitoring stations collecting precipitation data, with only two of them additionally measuring temperature. However, the Foum Zaabel station stands out for its ability to measure other crucial parameters (Bouziane et al. 2024; El Ouali et al. 2023).

Recognizing the distinct capabilities of these stations, we strategically designated the Foum Zaabel station as the reference point for all subsequent analyses, ensuring a consistent and standardized approach to our dataset (Bouziane et al. 2024).

Given the variation in the monitored parameters and instrumentation across stations, the collected data arrives in different forms, introducing a layer of complexity to our analysis. That necessitates a meticulous pre-processing phase, which we will detail in a subsequent section. As highlighted earlier in our paper, we possess a dataset of historical records of the five meteorological features: precipitation, humidity, wind, temperature, and evaporation—sourced exclusively from the designed station, Foum Zaabel (Bouziane et al. 2024). The integration of data from multiple stations and periods enriches the dataset, offering a comprehensive perspective on the meteorological conditions within the Ziz Basin, despite the inherent variations in monitoring capabilities across stations.

To provide a visual insight into the diversity of our data sources, we present an illustrative image showcasing an example of the original form of temperature measurements collected from one of the monitoring stations in the Ziz Basin. Figure 2 is a snapshot capturing the raw and unprocessed nature of the monthly temperature data directly acquired from the field.

As depicted, the temperature measurements exhibit fluctuations and variations over time, reflecting the dynamic nature of climatic conditions within the study area. The visual representation of the original temperature data not only underscores the

Fig. 2 Monthly temperature measurements in the original form collected from one of the monitoring stations in the Ziz Basin

YEAR	September	October	November	December	January	February \
82/83	24.2	18.300000	12.300000	9.200000	6.900000	10.6
83/84	24.6	19.000000	15.300000	9.400000	7.500000	10.6
84/85	24.4	18.100000	12.200000	8.700000	7.000000	12.3
85/86	24.1	18.500000	13.900000	8.600000	9.200000	11.4
86/87	24.9	16.900000	12.400000	8.760000	10.500000	11.9
87/88	25.6	19.400000	12.600000	10.200000	9.400000	11
88/89	24.3	18.500000	12.800000	6.900000	5.500000	8.9
89/90	25.1	19.000000	13.200000	10.800000	7.600000	12.8
90/91	24.6	19.800000	13.900000	7.600000	6.500000	7.8
91/92	24.4	16.700000	12.500000	9.800000	5.700000	9
92/93	24.2	17.900000	12.800000	8.800000	6.200000	8.2
93/94	*	14.900000	11.300000	9.800000	8.600000	9.9
94/95	23.5	17.000000	11.800000	8.700000	8.100000	11
95/96	*	17.200000	14.300000	10.000000	9.900000	9.9
96/97	23.7	19.300000	13.300000	9.800000	9.300000	12.5
97/98	22.8	19.500000	14.300000	9.500000	8.700000	9.9
98/99	25.4	18.200000	13.500000	7.300000	7.200000	9.2
99/2000	24.7	19.000000	10.600000	7.800000	5.500000	11.3
2000/2001	23.8	16.400000	18.400000	9.500000	8.800000	9.3
2001/2002	25.1	21.150000	12.800000	11.900000	9.100000	14.3
2002/2003	24.5	19.500000	13.500000	10.500000	8.700000	15
2003/2004	25.4	19.300000	12.700000	10.300000	10.200000	10.8
2004/2005	24.8	20.500000	11.400000	7.700000	6.700000	*
2005/2006	*	20.000000	12.500000	8.050000	5.800000	9.2
2006/2007	24.1	20.200000	13.400000	7.130000	8.200000	12.2
2007/2008	24.8	18.000000	12.700000	8.600000	8.200000	9.4
2008/2009	23.7	15.300000	9.600000	7.200000	7.800000	10.2
2009-2010	6.321667	12.340323	14.503333	11.496774	11.477419	13.583929
2010-2011	6.448333	10.051613	14.275000	11.403226	9.306452	10.669643
2011-2012	6.291667	10.608065	12.893333	10.011290	7.929032	8.010345
2012-2013	5.896667	10.020968	13.810000	8.787097	9.354839	9.808621
2013-2014	5.753333	11.658065	11.681667	7.003226	9.533871	11.52069
2014-2015	6.488333	11.285484	11.868333	8.219355	7.588710	8.756897
2015-2016	*	NaN	NaN	NaN	NaN	NaN

diverse temporal periods covered but also emphasizes the need for pre-processing to extract meaningful patterns and trends.

In the upcoming subsection, we will delve into the specific pre-processing techniques applied to transform this raw data into a format conducive to robust analysis and accurate precipitation prediction within the study area.

6.3 Pre-processing Techniques

In this subsection, the focus now shifts to the critical stage of pre-processing techniques. In the following, we will explore specific methodologies employed to transform the raw data, ensuring it aligns with the prerequisites for robust analysis and precise precipitation prediction within the study area. These techniques address temporal and spatial disparities, enhance data integrity, and prepare for a nuanced exploration of meteorological intricacies within the Ziz Basin.

6.3.1 Restructuring Data

Data restructuring is a fundamental pre-processing technique tailored to optimize the organization and granularity of our meteorological information. Given the variability in daily measurements, especially for temperature and precipitation, we undertook a strategic reorganization to enhance the data's analytical utility.

One facet of restructuring involves temporal aggregation. We condense the data by transforming daily measurements into a monthly format while preserving essential temporal trends. This process streamlines the dataset and facilitates a more comprehensive understanding of variations over broader periods in the featured meteorological parameters.

As illustrated in Fig. 3, this technique aligns all features to a consistent monthly timeline, providing a structured foundation for subsequent analyses. It ensures that variations in precipitation, humidity, wind, temperature, and evaporation are collectively captured on a monthly scale, contributing to a holistic understanding of meteorological dynamics within the Ziz Basin.

6.3.2 Handling Missing Data: Imputation Technique

The integrity of meteorological datasets often faces challenges due to missing values, which can significantly impact the accuracy of analyses. To mitigate this, we employ imputation techniques as a crucial pre-processing step.

Imputation methods, such as mean imputation, forward-fill, or machine learning-based imputation, play a pivotal role. These techniques are adept at replacing missing values with estimated or predicted values, ensuring the temporal sequence remains intact.

```
sdf=pd.read_excel('Date_File.xlsx', parse_dates =True)
sdf.set_index('DATE')
#Monthly resample
data_sdf=sdf.resample("M", on='DATE').sum()
data_sdf=data_sdf.interpolate(method = 'linear')
print(data_sdf)
```

```
            PRECIPITATION
DATE
1982-09-30            0.0
1982-10-31            0.0
1982-11-30            0.0
1982-12-31            0.7
1983-01-31            0.2
...                   ...
2013-04-30            0.0
2013-05-31            0.0
2013-06-30            0.0
2013-07-31            0.0
2013-08-31            0.0
```

Fig. 3 Monthly aggregation of precipitation data

Figure 4 exemplifies how strategic filling of missing values enhances the completeness and reliability of our meteorological dataset. This process ensures a more accurate analysis in the study area, providing a solid foundation for reliable insights.

```
# Load the Excel file into a DataFrame
sdf = pd.read_excel('Evaporation_Date.xlsx', index_col=0)
sdf = sdf.replace('*', np.nan)
sdf['Evaporation'] = sdf['Evaporation'].astype(float)
print(sdf)
#Fillna with the mean for the same periode of the year
sdf = sdf.fillna(sdf.groupby(sdf.index.month).transform('mean').applymap(lambda x: '{:.2f}'.format(x)))

print(sdf)
sdf.to_excel("Evaporation_Date.xlsx")
```

```
            Evaporation
Date
1982-09-01        456.3
1982-10-01        330.7
1982-11-01        242.5
1982-12-01        163.0
1983-01-01        143.4
...                 ...
```

Fig. 4 A visual representation of imputation techniques addressing missing data in our meteorological dataset

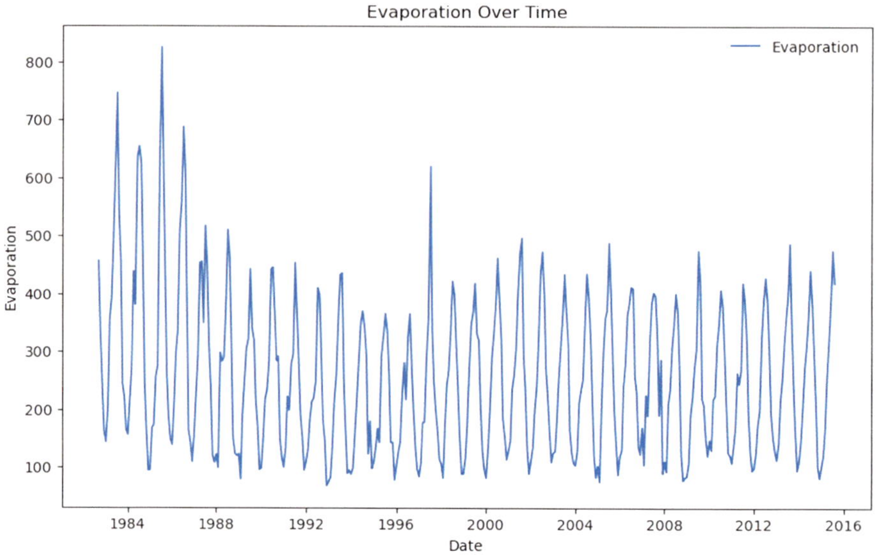

Fig. 5 Outlier detection: unveiling anomalies in evaporation

6.3.3 Identifying and Removing Data Outliers: Statistical Approaches

Detecting and handling outliers is crucial for maintaining the integrity of meteorological data. We use statistical methods like the interquartile range (IQR) or Z-score to identify these outliers systematically. Once identified, outliers are carefully removed to prevent them from influencing subsequent analyses.

Figure 5 examines evaporation trends over time. This visual insight aids in spotting outliers, ensuring the reliability of our meteorological dataset.

6.3.4 Feature Scaling: Standardization for Consistent Comparisons

Making sure different meteorological features are comparable is essential for accurate analysis. Feature scaling achieves this by adjusting the range of values across various features.

The standardization process transforms data to fit within a standardized range. Commonly used methods include z-score normalization or Min–Max Scaling. This method ensures that each feature contributes proportionally to the analysis, preventing the dominance of features with larger values.

Figure 6 illustrates the application of the Min–Max Scaling method to meteorological features. This technique ensures standardized ranges for variables like temperature, precipitation, humidity, wind, and evaporation, fostering scientific consistency and facilitating precise comparisons within the dynamic meteorological context of the study area.

```
# Load the dataset
df = pd.read_excel('Assembled_file.xlsx', sheet_name='1996-2015', parse_dates=['Date'], index_col='Date')

# Split the dataset into features (X) and target (y)
X = df.drop('Precipitation', axis=1)
y = df['Precipitation']

# Split the data into training and testing sets
train_size = int(len(df) * 0.7)
X_train, X_test = X[:train_size], X[train_size:]
y_train, y_test = y[:train_size], y[train_size:]

# Normalize the data
scaler_X = MinMaxScaler(feature_range=(0, 1))
scaler_y = MinMaxScaler(feature_range=(0, 1))
train_X = scaler_X.fit_transform(X_train)
train_y = scaler_y.fit_transform(y_train.values.reshape(-1, 1))
test_X = scaler_X.transform(X_test)
test_y = scaler_y.transform(y_test.values.reshape(-1, 1))

# Reshape the data to fit the LSTM input shape
train_X = train_X.reshape((train_X.shape[0], 1, train_X.shape[1]))
test_X = test_X.reshape((test_X.shape[0], 1, test_X.shape[1]))
```

Fig. 6 Harmonizing meteorological variables: Min–Max scaling for scientific consistency

```
# Make predictions on the test data
y_pred = model.predict(test_X)

# Inverse transform the predictions and actual values back to the original scale
y_pred = scaler_y.inverse_transform(y_pred)
y_test = scaler_y.inverse_transform(test_y)
```

Fig. 7 Illustration of the inverse transformation process applied to predicted results

After the meteorological data undergoes pre-processing and serves as input for predictions, a crucial step is the inverse transformation of the results. This process utilizes the inverse of the applied scaling method to revert the predicted outcomes to their original scale (Fig. 7).

The significance of this step lies in ensuring that the predicted results align seamlessly with the units and context of the initial meteorological data, providing a coherent and interpretable outcome.

After carefully exploiting the previous pre-processing techniques, the following subsection will explore the outcomes of these steps.

6.4 Results After Pre-processing

As shown in Fig. 8, we have meticulously refined the dataset, which currently covers the period of September 1996 to August 2015 with 228 records, showcasing our commitment to detailed pre-processing.

```
        Date  Evaporation  Wind  Humidity  Temperature  Precipitation
0 1996-09-01       271.19  2.38     27.06         23.7            3.5
1 1996-10-01       201.82  1.92     31.88         19.3            3.0
2 1996-11-01       135.90  1.98     29.00         13.3            0.0
3 1996-12-01        97.28  2.30     42.00          9.8           17.8
4 1997-01-01        84.66  2.35     40.00          9.3           27.9
<class 'pandas.core.frame.DataFrame'>
RangeIndex: 228 entries, 0 to 227
Data columns (total 6 columns):
 #   Column         Non-Null Count  Dtype
---  ------         --------------  -----
 0   Date           228 non-null    datetime64[ns]
 1   Evaporation    228 non-null    float64
 2   Wind           228 non-null    float64
 3   Humidity       228 non-null    float64
 4   Temperature    228 non-null    float64
 5   Precipitation  228 non-null    float64
dtypes: datetime64[ns](1), float64(5)
```

Fig. 8 Summary of the refined dataset providing a total of 228 records of meteorological insights

Our focus on outlier handling has successfully removed irregularities, ensuring clarity and reliability. Additionally, we addressed null data points and missing values using interpolation techniques, resulting in a complete and continuous dataset.

Figure 9 provides a consolidated overview of essential metrics (including counts, means, standard deviations, minimums, and maximums) for the five meteorological parameters. This snapshot equips researchers with a brief understanding of the dataset's distribution and sets the stage for a more nuanced exploration of weather patterns in the study area.

In addition, understanding the interactions between precipitation, humidity, wind, temperature, and evaporation is essential for unravelling the complex climate patterns in this area. Figure 10 sheds light on these correlations, offering a visual representation that lays the groundwork for a deeper exploration of interconnected meteorological phenomena.

Through a scatterplot matrix, the visualization unveils how parameters such as wind, evaporation, humidity, temperature, and precipitation interact throughout the

	Evaporation	Wind	Humidity	Temperature	Precipitation
count	228.000000	228.000000	228.000000	228.000000	228.000000
mean	244.366342	2.354649	34.997412	18.563640	18.469795
std	117.441752	0.406610	10.899375	7.603158	29.343143
min	75.500000	1.250000	11.000000	5.500000	0.000000
25%	133.020000	2.080000	28.560000	11.985000	0.600000
50%	232.285000	2.380000	33.000000	18.520000	6.400000
75%	336.040000	2.632500	41.000000	25.407500	23.500000
max	620.060000	3.420000	72.000000	31.140000	180.800000

Fig. 9 Summary statistics of the five meteorological parameters

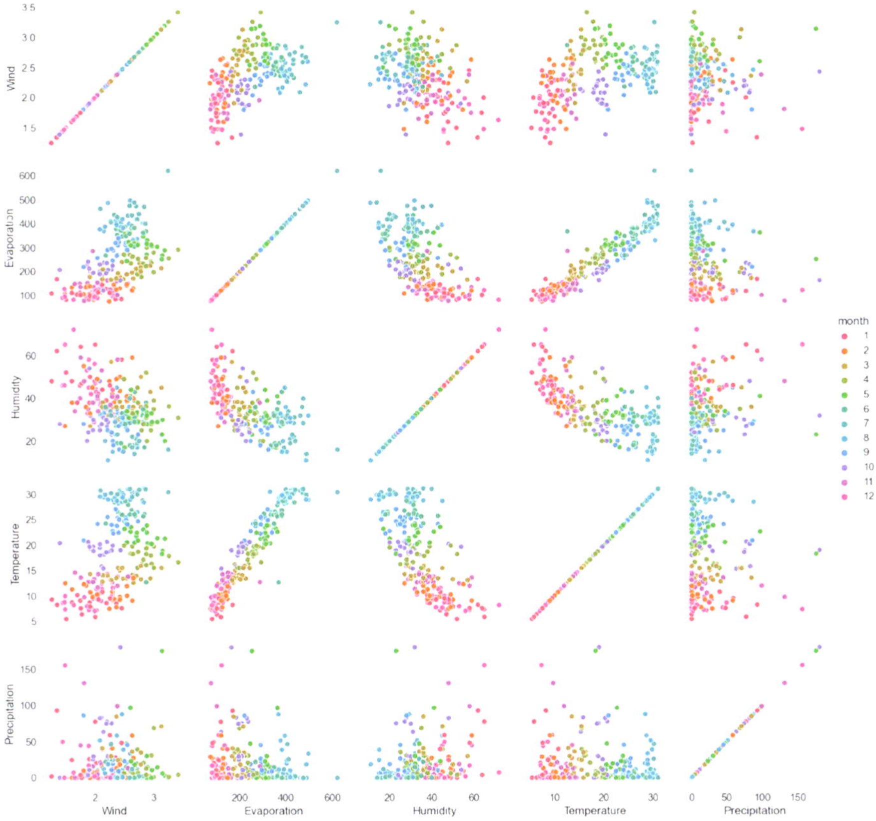

Fig. 10 Interplay between meteorological factors across different months, providing insights into their pairwise relationships (Bouziane et al. 2024)

year (Bouziane et al. 2024). The varying shades in the matrix denote the progression of months, facilitating the identification of any noticeable tendencies or periodical habits in the correlations among these elements within the study area (Bouziane et al. 2024).

With the preceding enhancements, the dataset is now ready for accurate analyses, providing a comprehensive overview of meteorological conditions in the Ziz Basin over nearly two decades.

Also, integrating modern approaches with this meticulously processed data is instrumental in advancing the accuracy and reliability of precipitation predictions, offering valuable insights for effective water resource management and informed decision-making in this study area.

This exploration aims to enrich the existing knowledge base and lay the groundwork for refining our approach and potential improvements in rainfall forecasting within our study area. It is a stepping stone for future research, fostering ongoing advancements in our understanding and capabilities in meteorological prediction.

7 Conclusion and Perspectives

Precise rainfall prediction is crucial for various sectors, ranging from agriculture to catastrophe risk reduction, directly influencing the decision-making process and resource allocation, which affects the overall resilience and preparedness of communities and ecosystems (Bouziane et al. 2024). However, it is essential to recognize the existing challenges within rainfall prediction, such as the inherent complexity of meteorological patterns and the uncertainties posed by climate change. Addressing these challenges is crucial for advancing our ability to provide reliable and timely predictions.

Traditional approaches exhibit limitations in addressing the complexity and non-linearity inherent in rainfall prediction. In contrast, Machine Learning techniques, such as Artificial Neural Networks, Support Vector Machines, and Long Short-Term Memory networks, offer promising avenues due to their adaptability to intricate meteorological data, promising to enhance predictive capabilities significantly (Bouziane et al. 2024).

In the specific context of the Ziz basin, these applications transcend weather prediction, impacting water resource management, agriculture, and disaster menace mitigation (Bouziane et al. 2024).

As a perspective, our objective is to construct an enhanced rainfall prediction model tailored to the designed study area, leveraging the meticulously processed and refined dataset from this region, which undergoes thorough pre-processing for data accuracy and reliability. Subsequently, we aim to analyze and compare the outcomes produced by these refined techniques with the results obtained from the traditional methods previously examined (Bouziane et al. 2024).

References

Al Mamun A et al (2018) Estimation of short-duration rainfall intensity from daily rainfall values in Klang Valley, Malaysia. Appl Water Sci 8:203. https://doi.org/10.1007/s13201-018-0854-z

AlHassoun SA (2011) Developing an empirical formulae to estimate rainfall intensity in Riyadh region. J King Saud Univ Eng Sci 23:81–88. https://doi.org/10.1016/j.jksues.2011.03.003

Barrera-Animas AY et al (2022) Rainfall prediction: a comparative analysis of modern machine learning algorithms for time-series forecasting. Mach Learn Appl 7:100204. https://doi.org/10.1016/j.mlwa.2021.100204

Bouziane S et al (2021) Proposal of a big data system for an intelligent management of water resources. In: Proceedings of the 6th edition of SAWIS international conference on water management, applied computing and data science (WMAD21), E3S web conference, vol 314, p 02002. https://doi.org/10.1051/e3sconf/202131402002

Bouziane S et al (2022) The use of big data in water resources management. In: Proceedings of the 2nd international conference on big data, modelling and machine learning (BML). SciTePress, pp 38–44. https://doi.org/10.5220/0010728000003101

Bouziane S et al (2024) Enhancing precipitation prediction in the Ziz basin: a comprehensive review of traditional and machine learning approaches, In: Proceedings of the 4th international GIRE3D

congress, E3S web of conferences, vol 489, p 04010. https://doi.org/10.1051/e3sconf/202448904010
Chao Z et al (2018) Research on real-time local rainfall prediction based on MEMS sensors. J Sens 2018:1–9. https://doi.org/10.1155/2018/6184713
El Ouali A et al (2023) Hydrological modeling and impact of climate change on water resources in the Ziz Valley, Central High Atlas, Morocco. Ecol Eng Environ Technol 24(6):192–210. https://doi.org/10.12912/27197050/168335
Evgeniou T, Pontil M (2001) Support vector machines: theory and applications. Lecture notes in computer science, pp 249–257. https://doi.org/10.1007/3-540-44673-7_12
Heistermann M et al (2013) Technical note: an open source library for processing weather radar data (wradlib). Hydrol Earth Syst Sci 17:863–871. https://doi.org/10.5194/hess-17-863-2013
Lynch P (2007) The origins of computer weather prediction and climate modelling. J Comput Phys 227:3431–3444. https://doi.org/10.1016/j.jcp.2007.02.034
Mehdaoui R et al (2018) Caractérisation à l'aide du SPI de la Sécheresse climatique dans le bassin versant de Ziz (Sud-Est, Maroc). Eur Sci J 14(21):177. https://doi.org/10.19044/esj.2018.v14n21p177
Narayanakumar S, Raja K (2016) A BP artificial neural network model for earthquake magnitude prediction in Himalayas, India. Circuits Syst 07(11):3456–3468. https://doi.org/10.4236/cs.2016.711294
Niu M et al (2016) An innovative integrated model using the singular spectrum analysis and nonlinear multi-layer perceptron network optimized by hybrid intelligent algorithm for short-term load forecasting. Appl Math Model 40:4079–4093. https://doi.org/10.1016/j.apm.2015.11.030
Ojo OS et al (2022) Machine learning models for prediction of rainfall over Nigeria. Sci Afr 16:e01246. https://doi.org/10.1016/j.sciaf.2022.e01246
Pham BT et al (2020) Development of advanced artificial intelligence models for daily rainfall prediction. Atmos Res 237:104845. https://doi.org/10.1016/j.atmosres.2020.104845
Villarini G et al (2014) Spatial and temporal modeling of radar rainfall uncertainties. Atmos Res 135–136:91–101. https://doi.org/10.1016/j.atmosres.2013.09.007
Xiang Y et al (2018) A SVR–ANN combined model based on ensemble EMD for rainfall prediction. Appl Soft Comput 73:874–883. https://doi.org/10.1016/j.asoc.2018.09.018
Yucel I, Onen A, Yilmaz KK, Gochis DJ (2015) Calibration and evaluation of a flood forecasting system: utility of numerical weather prediction model, data assimilation and satellite-based rainfall. J Hydrol 523:49–66. https://doi.org/10.1016/j.jhydrol.2015.01.042

Evaluation of the Effects of Climate Change on Surface Water Resources at the Ain Kwachia Dam Using Machine Learning Between 2008 and 2021

Mohamed Gramz, Mouhcine Batchi, Moulay Hicham Azagane, Adnane El-Boukhari, Mehdi Mettouchi, and Jamal El Bouziani

Abstract Surface water resources are subject to climate change, particularly in arid zones such as Morocco's central plateau, and to the massive demand placed on water by human activities, which calls for equitable management of water potential, In this sense, hydraulic engineering is a solution for better storage and management of surface water, as well as for mitigating and adapting to climate variability. This is the aim of the Ain Kwaachia dam on the Ykem river in the northwestern part of the central plateau of Morocco. While the use of remote sensing technology, based on the acquisition of satellite images, is a kind of machine learning to quantify spatio-temporal variations in water potential, so the processing and the analysis of these image data, based on a GIS geographic information system, and this, from the Land Use index for land cover, with an unsupervised classification from 2008, the year of commissioning of the dam subject of this research, and a supervised classification around the year 2021. The results obtained show that the variations affecting the study area have been noticeable in recent years, from 9.78 km^2 for 2016, to 4.23 km^2 for 2021, as a result of the repercussions of climate change, with falling rainfall on the one hand, and in the other hand rising temperatures.

Keywords Water resources · Machine learning · GIS · Land use · Ain Kouachia dam

M. Gramz (✉) · M. Batchi · M. H. Azagane
Ibn Tofail University, Territory, Environment and Development Laboratory, Kenitra, Morocco
e-mail: mohamed.gramz@uit.ac.ma

A. El-Boukhari
Ibn Tofail University, Literature, Arts and Pedagogical Engineering Laboratory, Kenitra, Morocco

M. Mettouchi
Hassan II University, Space, Planning and Environment Laboratory, Casablanca, Morocco

J. El Bouziani
National Department of Planning Territory, Rabat, Morocco

N. Rebai et al. (eds.), *GIS, Applied Computing and Data Science for Water Management*, Lecture Notes in Geoinformation and Cartography,
https://doi.org/10.1007/978-3-031-63038-5_13

1 Introduction

Morocco's geographical location at latitudes between 21° and 36° north of the globe's northern hemisphere characterizes the country's climate, which is marked by aridity over most of its territory, with a reduction in inter-annual rainfall and an increase in temperatures (Iouzzi et al. 2023), sometimes leading to problems of water scarcity in certain regions of the country. It is therefore necessary to take short- and medium-term mitigation measures and adopt long-term adaptation strategies in order to cope with extreme phenomena linked to climate change (Dallahi et al. 2020; Mehboob and Kim 2021). With this in mind, the Moroccan authorities in charge of water, have been building dams for many years, in order to better manage and store water resources, and to ensure the availability and sustainability of this natural potential resource, to meet the needs of the population and its activities (agricultural, industrial), on the one hand, and in the other hand to safeguard environmental balances and eco-systemic stability. It is in this sense that our research focuses on surface water variability at the Ain Kwachia dam, located on the Ykem river in the northwest of the central plateau of Morocco (see Fig. 1) over the period 2008 to 2021, using a methodological and operational approach based on remote sensing data and machine learning techniques, initially to estimate the water surfaces at the dam, in terms of monitoring the evolution of surface water resources in this hydraulic structure, via artificial intelligence technology, and secondly to give visibility and help water managers predict future water resource volumes, to anticipate likely scenarios in the event of climatic hazards, i.e. drought, floods, in order to make the best decisions on water management policy, in this geographical location. The usefulness of machine learning technology, as an artificial intelligence solution that is illustrated by the acquisition of satellite images, and their processing by a GIS geographic information system (Islam et al. 2021; Lahcen et al. 2022), these geospatialization tools (Karim et al. 2019; Saadi et al. 2021; Pádua et al. 2022; Parveen and Ilahi 2022; Revuelta-Acosta et al. 2022), as effective tools for monitoring natural phenomena, thus overcoming the constraints of data scarcity over time, help to map surface water in the study area, and of course at the Ain Kwachia dam, with validation of the results for the current situation by location visits of this hydraulic structure, our case study for this article.

The methodology of our work is based on the use of the Land Use index to quantify spatio-temporal variations in water resources (Bouizrou et al. 2022; Amiri et al. 2022), in this geographical area of Morocco. This analysis method is a complementary approach for assessing surface water variation (Leya et al. 2022), and thus for solving water resources related problems. The quantification of water potential in the low part of the central Moroccan plateau, our study area, is characterized by a mediterranean climate (Brouziyne et al. 2020), its geographical location in the north-west, topographically limited to the north by the Bouragrag watershed, to the south by the low chaouia, to the east by the intermediate plateau, and to the west by the Atlantic Ocean, with altitudes varying between −8 m and 465 m from the coastal part to the mainland (see Fig. 1). Average annual rainfall over recent years (2008–2021) is around 428 mm, according to values recorded by rainfall stations. It is characterized by monthly and daily irregularity, with the exception of 2010 when rainfall reached 864.3 mm, the highest value in the data series used, at the Cheikh reguig rainfall station for the Zaer plateau. This climatic situation, together with morphological characteristics, has an impact on surface water resources.

In general, the majority of studies on water potential for the Moroccan central plateau are based on the vast Bouragrag basin, which spans a large area. This means that additional specific research is needed for geographic areas where the need for water resources is a priority in order to better manage surface water potential through the construction of hydraulic works with the goal of both adapting to water scarcity and mitigating the short- and medium-term effects of climate change. The purpose of this research is to address the extreme case of climate change and plan for the conservation and equitable management of water resources by using artificial intelligence technology to assist water managers in the study area with surface water management.

2 Materiel and Method

Quantifying the values of water resource variability requires, in the absence of statistics and weak data, the use of a performant tool or index to monitor the spatio-temporal evolution of surface water resources, as is the case for the Land Use index (Ikraoun et al. 2021). Processing these results gives us measurable values, and enables us to design any kind of variation in water potential (Elassassi et al. 2022), in time and space.

3 Study Area

The Ain Kwachia dam is located on the Ykem river, in the low part of the central plateau in Morocco, and is one of the main sources of water supply, especially for human activities on the Zaer plateau. It covers an area of 124 ha, in the Sidi Yahya Zaer local authority, and includes the adjacent Ykem and Cherrat basins, bounded to the north by the Bouragrag basin (Khomsi et al. 2016), to the south by low chaouia, to the east by the intermediate plateau, and to the west by the Atlantic Ocean, the topography of this area ranges from −8 m to 60 m in the coastal part, where dunes are designated, to heights of up to 250 m in the middle, covering half the surface area, while the remainder reaches 450 m, especially in the south-west, with a maximum altitude value of 464 m (see Fig. 1). In climatic terms, this is a semi-arid region (El Kasri et al. 2021; Tuel and Eltahir 2018) where the average annual rainfall is almost 456 mm. The average temperature varies between 18 and 23 °C, with variability between the summer and winter seasons. In addition, this area is characterized by vegetation cover in the form of forests located to the west mostly cork oak (Dallahi et al. 2023).

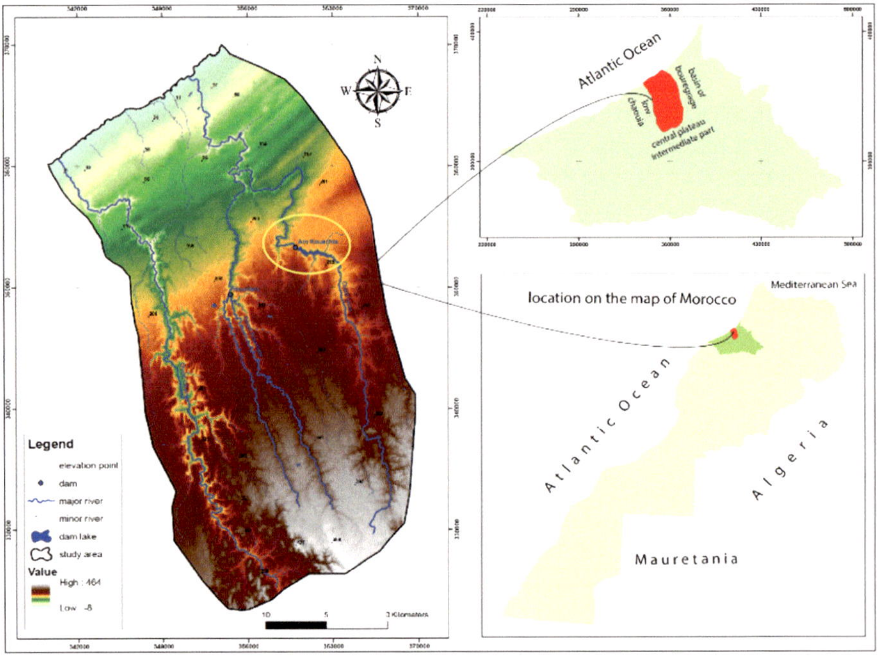

Fig. 1 Localization of the study area. *Source* Based on the topographic maps of Temara, Benslimane and Sidi Battache plus the digital elevation model DEM

4 Methodology

In this article, our methodological approach to monitor the evolution of surface water resources at the Ain Kwachia dam, located in the north-western part of Morocco's central plateau (see Fig. 1), over the period 2008–2021, is based on monitoring the historical spatial evolution of this area on a more temporal scale, through the analysis of satellite images via a GIS geographic information system, according to the Land Use index for land occupation, in the first instance, then mapping the variation trends, the results obtained being explained towards the end, by an interpretation of the climatic data characterizing the target zone.

5 Land Use

The advantage of using the Land Use index (Kenawi et al. 2023) as a GIS-based machine-learning technique is that it quantifies land cover in terms of its various components, with a high degree of precision, for the purpose of assessing spatio-temporal variations. This is not possible with other indices that deal with a single

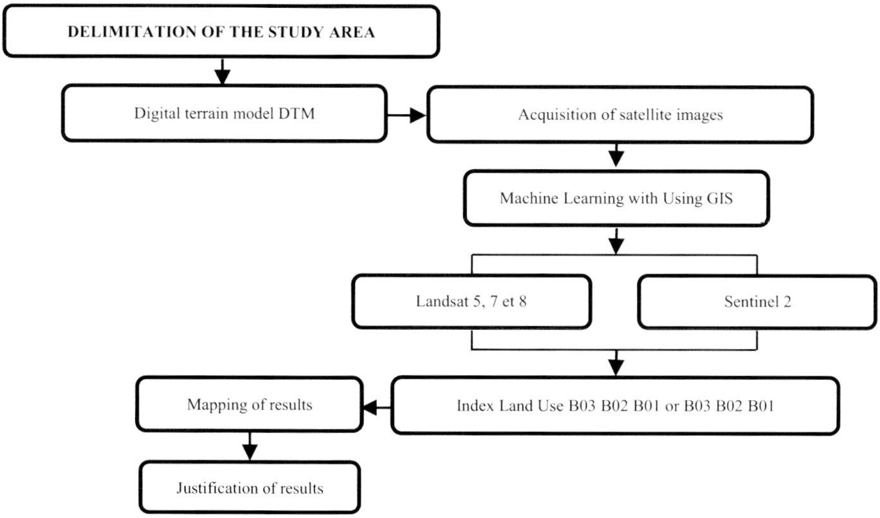

Fig. 2 Framework of the methodological approach for this research

component, such as the Normalized Difference Vegetation Index NDVI and the Normalized Difference Water Index NDWI.

Using GIS, the Land Use index enables us to analyze the types of land use in the geographical area of our study. This index is created by combining several layers of data, such as land cover, vegetation, water, roads and urban areas. As a first step, we combined the bands that give the true image of the Earth (natural color), which differs from one satellite to another. Between 2008 and 2021, we used a combination of bands for each satellite image, as shown in the Table 1.

Table 1 Satellite images used between 2008 and 2021 for land use

Year	Satellite image	Band used
2008	Landsat 5	B03 B02 B01
2011	Landsat 7	B03 B02 B01
2016	Landsat 8	B04 B03 B02
2021	Sentinel 2	B04 B03 B02

Source https://www.usgs.gov/

In this table, the satellite images available for 2008 and 2011 are Landsat 5 and 7 respectively, with spectral bands B03, B02 and B01. For 2016 and 2021, the bands used are B04, B03 and B02, with satellite images from Landsat 8 and Sentinel 2.

In the second stage, we use an -image classification- tool to draw each type separately (water, matorral, bare land, agriculture). Once we've finished drawing samples of the above-mentioned types, we perform a supervised classification in the third step, using an -interactive supervised classification- tool, which gives us the land uses in our study area.

6 Data

The data used to interpret the results are based on precipitation and temperature values, recorded respectively by the rainfall stations provided by the state services in the case of the agency for the lower hydraulic basin of Bouragrag and the Chaouia ABHBC, and meteorological values, in terms of temperature, collected from the NASA website (Guo et al. 2019).

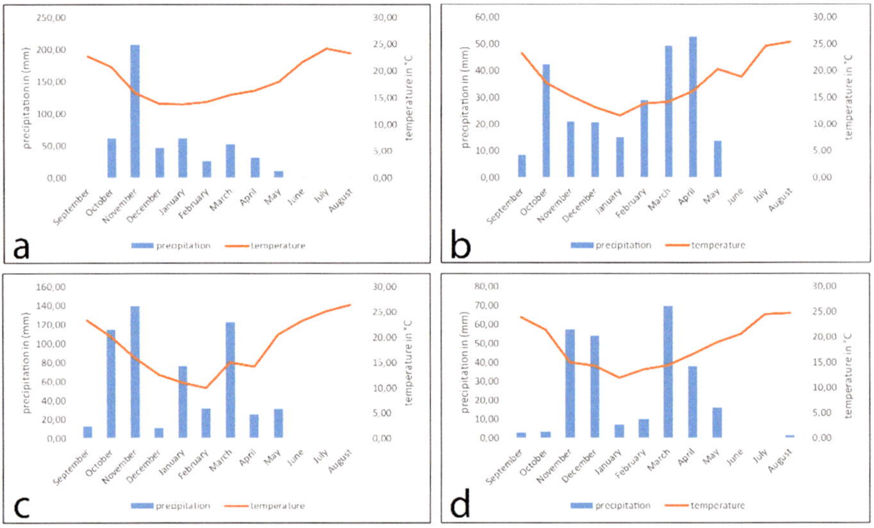

Fig. 3 Monthly values of precipitation and temperature recorded at rainfall and weather stations respectively **a** 2008, **b** 2011, **c** 2016, **d** 2021. *Source* ABHBC data and NASA website https://power.larc.nasa.gov/data-access-viewer/

7 Results and Discussion

The analysis carried out as part of our research paper enable us to produce maps showing the evolution and variability of water resources, which were produced on the basis of the processing carried out during the four reference years 2008, 2011, 2016 and 2021 following the methodological approach (see Fig. 2).

With regard to the results of the Land Use index, over the period 2008 to 2021, the analysis results are divided into four categories, respectively classified in terms of surface area from large to small, bare land, matoral, agricultural land and water surface, marked by an increase in the percentage of agricultural land from 20.25 to 46.74 and water surface from 3.78 to 9.78, between the period 2008 and 2016, with a decrease between 2016 and 2021, with a percentage of 43.02 for agricultural land and 4.23 for water surface, while the values obtained for bare land decreased from 57.22% to 24.33% between 2008 and 2016, and increased to 32.03% towards the year 2021, for matoral surface, characterized by an increase from 22.09 to 32.58% between 2008 and 2011, on the contrary the decrease of the percentage to 24.46 between 2011 and 2021 (see Table 2 and Fig. 4).

Table 2 Values in km^2 of land area distribution between 2008 and 2021 based on land use

Year	Bare land in Km^2	Water in Km^2	Agricultural land in Km^2	Matoral in Km^2	Total surface in Km^2
2008	489,49	3,78	173,29	188,94	855,50
2011	358,14	6,66	211,93	278,77	855,50
2016	208,1	9,78	399,9	237,72	855,50
2021	274	4,23	368,01	209,26	855,50

Source Results of using geostatial techniques

The land use values shown in Table 2 illustrate the variation in the study area's constituents. The surface area of water, agricultural land, and matoral increased in km^2 between 2008 and 2016 and decreased between 2016 and 2021, while the surface area of bare land decreased in km^2 between 2008 and 2016 and increased toward the year 2021. The evolution of the surface area occupied by the Ain Kwachia dam is mapped with a zoom on the dam area, following its size in relation to the study area (El Bilali et al. 2022), for greater clarity on these spatiotemporal variations from 2008 to 2021 (see Figs. 5 and 6), This chronology shows an increase in the surface area of the Ain Kwachia dam.

To visualize the spatiotemporal variation in the study area, the land use representation is mapped (see Fig. 4) for the reference years as indicated in Table 2.

A change in the water surface area was observed (see Fig. 6) from 2008, the year the Ain Kwachia dam was put into service, to 2021, the last reference year for the investigation of the evolution of water resource storage, even though the total surface area under water decreased.

After applying the methodological approach adopted for this research, for the quantification of spatiotemporal variations in water resources, the evolution of the water surface is remarkable over the periods between 2008 and 2021 at the level of the Ain Kwachia dam (see Fig. 5) on the one hand, with a steady growth, and thus, per the findings of the analysis for the land use index used, which is justified by the precipitation data (see Fig. 3), where high rainfall was recorded between 2008 and 2011 during the winter season (December, January and February), the period

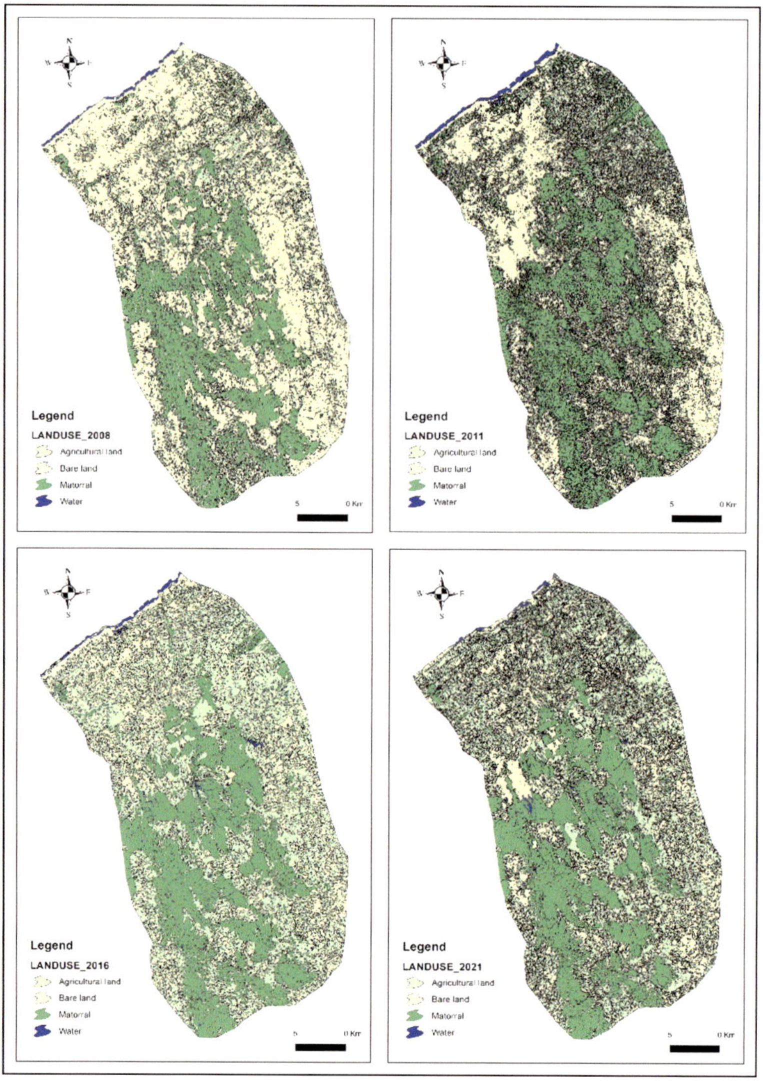

Fig. 4 Results using GIS for the land use index in the area study between 2008 and 2021. *Source* Results of using GIS system

between 2011 and 2016 is characterized by a rise in temperature, which favours evaporation and evapotranspiration of water, as well as a decline in precipitation (see Fig. 3), with the effect of climate change during this period (2021), but as far as surface water resources in the study area are concerned, we have a decrease between 2016 and 2021 from 9.78 to 4.23 km^2 (see Table 2), which explains the impact of climate change in terms of rainfall, despite the stability of temperatures throughout the year for the four reference years (see Fig. 3). The dominance of dry years has

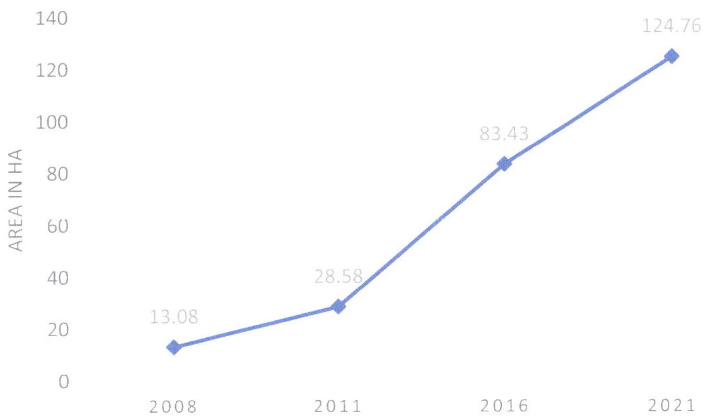

Fig. 5 Evolution of the area occupied by the Ain Kwachia dam between 2008 and 2021. *Source* Results of using geostatial techniques with data from ABHBC and NASA

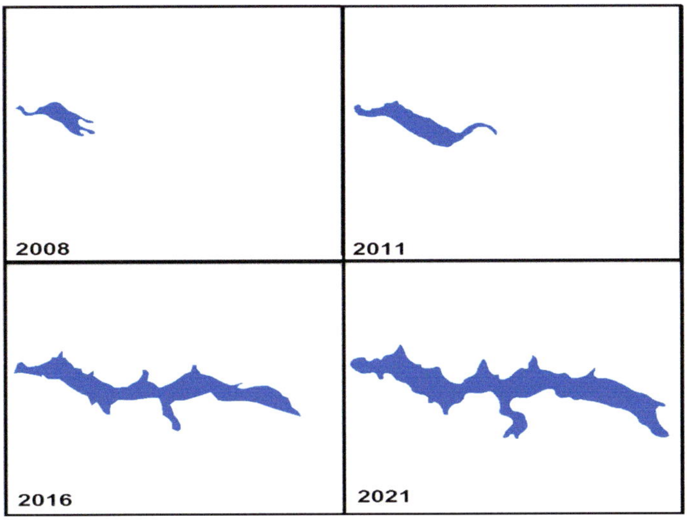

Fig. 6 Evolution of the area occupied by the Ain Kwachia dam between 2008 and 2021. *Source* Based on the use of a GIS tool

direct repercussions on surface water quantities, precisely following the results of Land Use, so the evolution of agricultural land between 2008 and 2016 from 173.29 to 399. 9 km^2, has an effect on surface water resources, in terms of consumption by anthropic activities, namely agriculture and poultry farming, from the principal rivers of the hydrographic network, with The dominance of dry years has direct repercussions on surface water quantities, precisely in line with the results of Land Use, so the evolution of agricultural land between 2008 and 2016 from 173.29 to

399.9 km^2, has an effect on surface water resources, in terms of consumption by anthropogenic activities, namely agriculture and poultry farming, from the main rivers of the local hydrographic network.

8 Conclusion

Currently, the use of artificial intelligence technology in combination with the functionalities of geospatialization tools such as remote sensing and GIS, has made it possible to track changes in the environment, and more specifically climate change as extreme phenomena that impact on natural potential, of which water is a part, to essentially meet the needs of the population of the sidi yahya zaer region in the northwestern central plateau of Morocco, the location of the Ain Kwachia dam, and of course to serve anthropic activities based on water resources, during all phases of the economic production chains, furthermore the results of this research aim to help territorial managers and dictators to take the necessary measures to preserve water on the one hand, and on the other to manage this water potential in an equitable and rational manner, in order to ensure the sustainability of water resources in the study area.

The specific climate of the Moroccan central plateau necessitates the use of additional hydraulic facilities to store surface water. One such facility is Ain Kwachia, which is essential to meeting the water needs of the local community despite the effects of climate change on rainfall in this region.

Acknowledgements Special thanks to the work team for the elaboration of this research, in addition to the data providers, namely the Bouragrag water basin agency and the Chaouia ABHBC, as well as NASA open access.

References

Amiri M, Salem A, Ghzal M (2022) Spatial-temporal water balance components estimation using integrated GIS-based Wetspass-M model in Moulouya Basin, Morocco. ISPRS Int J Geo-Inform 11:139. https://doi.org/10.3390/ijgi11020139

Bouizrou I, Aqnouy M, Bouadila A (2022) Spatio-temporal analysis of trends and variability in precipitation across Morocco: comparative analysis of recent and old non-parametric methods. J Afr Earth Sc 196:104691. https://doi.org/10.1016/j.jafrearsci.2022.104691

Brouziyne Y, Abouabdillah A, Chehbouni A et al (2020) Assessing hydrological vulnerability to future droughts in a Mediterranean watershed: combined indices-based and distributed modeling approaches. Water (Switzerland) 12:2333. https://doi.org/10.3390/W12092333

Dallahi Y, Boujraf A, Meliho M, Orlando CA (2023) Assessment of forest dieback on the Moroccan Central Plateau using spectral vegetation indices. J for Res 34:793–808. https://doi.org/10.1007/s11676-022-01525-x

Dallahi Y, Ouhammou A, Sbai M et al (2020) Assessment of the vegetation cover change impacts on water erosion, using pap/rac method in upstream of "ouljet soltane" dam, central plateau-Morocco. Geographia Technica 15:153–161. https://doi.org/10.21163/GT_2020.151.14

El Bilali A, Taleb I, Nafii A, Taleb A (2022) A practical probabilistic approach for simulating life loss in an urban area associated with a dam-break flood. Int J Dis Risk Reduct 76:103011. https://doi.org/10.1016/j.ijdrr.2022.103011

El Kasri J, Lahmili A, Soussi H et al (2021) Trend analysis of meteorological variables: rainfall and temperature. Civil Eng J (Iran) 7:1868–1879. https://doi.org/10.28991/cej-2021-03091765

Elassassi Z, Ougrad I, Bedoui I et al (2022) Spatial and temporal variations of the water quality of the Tiflet River, Province of Khemisset, Morocco. Water (Switzerland) 14:1829. https://doi.org/10.3390/w14121829

Guo X, Yang Y, Li Z et al (2019) Drought trend analysis based on the standardized precipitation-evapotranspiration index using NASA's earth exchange global daily downscaled projections, high spatial resolution coupled model intercomparison project phase 5 projections, and assessment of potential impacts on china's crop yield in the 21st century. Water (Switzerland) 11:2455. https://doi.org/10.3390/w11122455

Ikraoun H, El Mderssa M, Malki F et al (2021) Assessment of land-use dynamics effects on water erosion: case of the Bouregreg subwatershed upstream of the Tiddas Dam, Morroco. Egypt J Soil Sci 61:459–467. https://doi.org/10.21608/ejss.2022.110612.1483

Iouzzi N, Ben Meftah M, Haffane M et al (2023) Modeling of the fate and behaviors of an oil spill in the Azemmour River Estuary in Morocco. Water 15:1776. https://doi.org/10.3390/w15091776

Islam H, Abbasi H, Karam A et al (2021) Geospatial analysis of wetlands based on land use/land cover dynamics using remote sensing and GIS in Sindh, Pakistan. Sci Progress 104:003685042110261. https://doi.org/10.1177/00368504211026143

Karim M, Maanan M, Maanan M et al (2019) Assessment of water body change and sedimentation rate in Moulay Bousselham wetland, Morocco, using geospatial technologies. Int J Sedim Res 34:65–72. https://doi.org/10.1016/j.ijsrc.2018.08.007

Kenawi MS, Alfredsen K, Stürzer LS et al (2023) High-resolution mapping of land use changes in Norwegian hydropower systems. Renew Sustain Energy Rev 188:113798. https://doi.org/10.1016/j.rser.2023.113798

Khomsi K, Mahe G, Tramblay Y et al (2016) Regional impacts of global change: seasonal trends in extreme rainfall, run-off and temperature in two contrasting regions of Morocco. Nat Hazard 16:1079–1090. https://doi.org/10.5194/nhess-16-1079-2016

Lahcen D, Hafida N, Souad M et al (2022) Integration of remote sensing and GIS in the identification of the vegetation covers degradation of the korifla basin (NW of Central Morocco) between 1990 and 2018

Leya RS, Bala SK, Newton IH et al (2022) Water security assessment of a peri-urban area: a study in Singair Upazila of Manikganj district of Bangladesh. Environ Dev Sustain 24:14106–14129. https://doi.org/10.1007/s10668-021-02023-6

Mehboob MS, Kim Y (2021) Effect of climate and socioeconomic changes on future surface water availability from mountainous water sources in Pakistan's Upper Indus Basin. Science Total Environ 769:144820. https://doi.org/10.1016/j.scitotenv.2020.144820

Pádua L, Duarte L, Antão-Geraldes AM et al (2022) Spatio-temporal water hyacinth monitoring in the lower Mondego (Portugal) using remote sensing data. Plants 11:3465. https://doi.org/10.3390/plants11243465

Parveen MT, Ilahi RA (2022) Assessment of land-use change and its impact on the environment using GIS techniques: a case of Kolkata Municipal Corporation, West Bengal, India. GeoJournal 87:551–566. https://doi.org/10.1007/s10708-022-10581-z

Revuelta-Acosta JD, Guerrero-Luis ES, Terrazas-Rodriguez JE et al (2022) Application of remote sensing tools to assess the land use and land cover change in Coatzacoalcos, Veracruz, Mexico. Appl Sci 12:1882. https://doi.org/10.3390/app12041882

Saadi O, Nouayti N, Nouayti A et al (2021) Application of remote sensing data and geographic information system for identifying potential areas of groundwater storage in middle Moulouya Basin of Morocco. Groundw Sustain Dev 14:100639. https://doi.org/10.1016/j.gsd.2021.100639
Tuel A, Eltahir EAB (2018) Seasonal precipitation forecast over Morocco. Water Resour Res 54:9118–9130. https://doi.org/10.1029/2018WR022984

The Use of Artificial Intelligence to Optimise Water Resources: A Comprehensive Assessment

Fouad Dimane, Yahya El Hammoudani, Lahcen Benaabidate, Khadija Haboubi, Abdelhak Bourjila, Chaimae Benaissa, Iliass Achoukhi, Abdelaziz Touzani, and Hatim Faiz

Abstract The issue of water management is at the heart of debates in many international forums. To meet the growing demand for water due to the current global crisis, it is essential to focus more on water management techniques applied in different applications. Taking into account population density, it becomes imperative to implement intelligent water management mechanisms for efficient distribution, preservation of water quality, and its use for various purposes. Our work addresses several key areas required for efficient water management, including wastewater recycling, water distribution, rainwater harvesting and irrigation, using artificial intelligence (AI) models. The data required for these applications is unique and varies according to context, so it is crucial to use a versatile model or algorithm capable of providing solutions for all these cases. AI, deep learning and Internet of Things techniques can facilitate the design of an intelligent water management system for the sustainable use of natural resources. Our work explores different approaches to water management using AI, DL and IoT, drawing on case studies and statistical analysis to develop a successful water management system.

Keywords Artificial intelligence (AI) · Water management · Water management techniques · Water distribution · Wastewater recycling

---

F. Dimane · Y. El Hammoudani (✉) · K. Haboubi · A. Bourjila · I. Achoukhi · A. Touzani · H. Faiz
National School of Applied Sciences of Al-Hoceima, Department of Energy and Environmental Civil Engineering, Engineering Sciences and Applications Laboratory, Abdelmalek Essaâdi University, Tetouan, Morocco
e-mail: elhammoudani5@gmail.com

L. Benaabidate
Laboratory of Functional Ecology and Environment Engineering, University of Sidi Mohammed Ben Abdellah, 30500 Fez, Morocco

C. Benaissa
Faculty of Science and Technology of Tangier, Department of Earth Sciences, Geosciences Research Team on Natural Risks, Abdelmalek Essaâdi University, Tetouan, Morocco

N. Rebai et al. (eds.), *GIS, Applied Computing and Data Science for Water Management*, Lecture Notes in Geoinformation and Cartography,
https://doi.org/10.1007/978-3-031-63038-5_14

1 Introduction

Water, as a vital resource of the earth's biosphere, is at the heart of a global crisis of alarming proportions. The growing demand for water resources, an unavoidable consequence of rapid population growth and environmental mutations induced by factors such as climate change, is exerting considerable and unprecedented pressure on the world's water stocks (Berthet et al. 2021; Sharan et al. 2023). This global water crisis means that we urgently need to fundamentally rethink our approach to ensuring the sustainability of our water supply and conserving the intrinsic quality of this vital resource (Irfeey et al. 2023; Taher et al. 2022).

In this critical situation, the need for effective water management techniques has become an inescapable imperative (Sadoff and Muller 2009). Today's challenges call for a proactive, integrated response that goes beyond traditional approaches. This is where intelligent water management comes into its own as a promising solution, with innovative prospects for overcoming these unprecedentedly complex challenges (Rijsberman 2006; Hammoudani et al. 2019, 2021, 2023; Hammoudani and Dimane 2020, 2021).

Rapidly changing environmental constraints, accentuated by growing demand for water resources, call for a fundamental review of our water management methods (Kabote and John 2017). Traditional paradigms, often limited in their ability to respond to increasingly complex issues, require a transition to innovative and adaptive approaches. Intelligent water management is distinguished by its ability to integrate emerging technologies such as artificial intelligence, deep learning and the Internet of Things, opening up new perspectives for tackling these multifaceted challenges (Sun and Scanlon 2019; Javed et al. 2022; Dimane and Hammoudani 2021; Bourjila et al. 2021, 2023; Benaissa et al. 2020, 2022).

This article sets out to explore the possibilities offered by artificial intelligence (AI), deep learning (DL), and the Internet of Things (IoT) in the field of water management (Lowe et al. 2022; Mustapha et al. 2021; Elabdouni et al. 2022). By placing technological innovation at the heart of our thinking, we seek to demonstrate how these advances can make a significant contribution to solving the crucial issues linked to the availability, distribution and sustainable use of water (Bourjila et al. 2021, 2023; Krishnan et al. 2022; Elabdouni et al. 2020; Abdouni et al. 2021; Bouhout et al. 2023; Achoukhi et al. 2023).

Our aim is to propose innovative perspectives and concrete solutions to guide the transition towards smarter, more efficient and environmentally-friendly water management (Fig. 1) (Krishnan et al. 2022). By combining the strengths of AI, DL, and IoT, we aspire to chart a course towards a future where water management is synonymous with sustainability, resilience, and the preservation of water resources for generations to come (Mondejar et al. 2021).

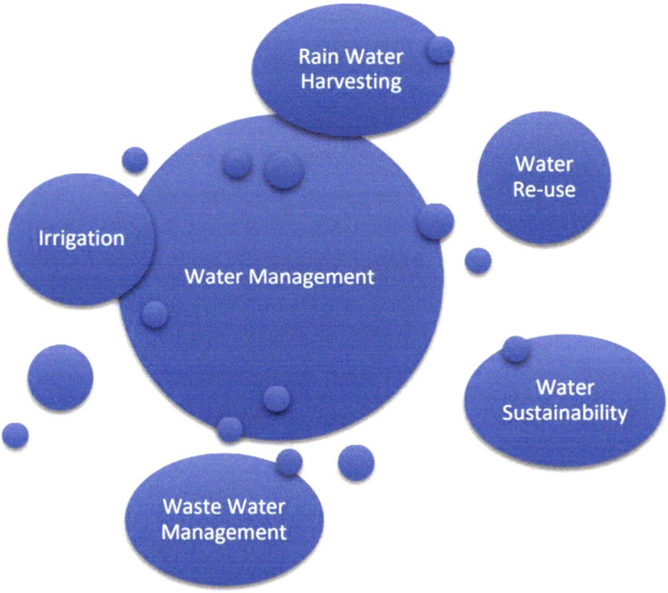

Fig. 1 Water management parameters (Krishnan et al. 2022)

2 Current Context of Global Water Management

2.1 The Current State of Global Water Management

On a global scale, water management remains a dynamic challenge, reflecting a complex and constantly evolving reality. Current trends highlight a diversity of challenges, from uneven availability to variable accessibility and quality (Schlueter et al. 2012). This complexity underscores the urgent need for flexible, adaptive approaches, capable of adjusting to specific regional contexts. Growing pressure on water resources stems from a number of factors, including rapid urbanization, increasingly marked climate change, and intensive agricultural practices that place significant demands on water reserves (Scott et al. 2021). These points of tension highlight the urgent need to rethink water management strategies on a global scale, integrating sustainable approaches and fostering international collaboration to meet these complex challenges (Pappalardo 2018). At the same time, significant advances in technology for monitoring and managing water resources offer promising opportunities for optimizing water use and alleviating pressure on these precious resources. As communities around the world strive to meet these challenges, water management is emerging as a crucial issue, underlining the importance of innovative policies and holistic solutions to ensure the long-term sustainability of water resources worldwide (Bogardi et al. 2012; Newson 2008; Fraiture et al. 2010).

2.2 The Role of Population Density in the Need for Intelligent Water Management

Population density is emerging as an inescapable catalyst for the growing complexity of water management requirements, particularly in densely populated urban areas. Exponential population growth within these areas creates a unique set of challenges, requiring a strategic and innovative approach to ensure sustainable water management (Brown et al. 2015; Zevenbergen et al. 2018).

One of the most immediate pressures resulting from population density is the dizzying increase in demand for water. Domestic, industrial and agricultural needs are intensifying, straining available water resources. What's more, this high concentration of people is putting a considerable strain on already aging water distribution infrastructures. Distribution networks are struggling to meet fluctuating and growing demand, creating gaps in water supply and increasing the risk of shortages (Zevenbergen et al. 2018).

At the same time, population density amplifies the risks of pollution, linked both to human activities and to the concentration of waste. Pressures on sanitation systems and the quality of water resources are reaching critical levels, requiring proactive measures to prevent adverse environmental impacts.

In the face of these challenges, intelligent water management is emerging as a promising solution. By thoroughly analyzing the specific pressures generated by high population density, it becomes possible to identify strategic opportunities for tackling these challenges in a targeted and efficient way. The integration of advanced monitoring technologies, for example, would enable predictive management of water demand, while the implementation of smart grids would optimize the distribution of water resources. In addition, initiatives focusing on sustainable practices would help reduce pollution risks (Subramani and Jacangelo 2015).

By taking a close look at these opportunities, we can lay the foundations for adaptive water management, capable of meeting the specific needs of densely populated urban areas. A holistic approach, combining technology, sustainability and proactive management, thus becomes imperative to ensure the resilience of water infrastructures and the preservation of water resources in a context of increasing urban densification (Eggimann et al. 2017).

2.3 Preserving Water Quality, Distribution Efficiency and Diversified Use

Preserving water quality and optimizing its distribution for diversified use are crucial elements in the current panorama of global water management. This duality poses major challenges that require innovative, integrated solutions to ensure the sustainability and continued availability of water resources (Hassan et al. 2021).

Persistent concern about water quality is rooted in the combined effects of industrial, agricultural and urban activities on a global scale (Bhateria and Jain 2016). Chemical discharges, agricultural contamination and industrial spills are all putting pressure on the intrinsic quality of our water supplies. These alterations not only threaten human health, but also have adverse repercussions on aquatic biodiversity and associated ecosystems.

At the same time, the efficient distribution of water to meet a variety of needs, from households to industry and agriculture, requires a global approach. Traditional distribution networks, prone to high losses and often inefficient, further complicate water management. The need for intelligent management is emerging as an essential response to these complexities (Zhou et al. 2019).

The adoption of innovative solutions, such as smart grids, advanced water quality monitoring and the integration of state-of-the-art treatment technologies, paves the way for optimizing distribution while preserving water quality. Technology plays a decisive role here, enabling early detection of contaminants, dynamic management of networks to minimize losses, and judicious allocation of water resources according to specific criteria (Wang et al. 2023).

Thus, the crucial importance of maintaining water quality while optimizing its distribution for diversified uses is a central challenge in water management worldwide. This duality calls for innovative approaches, harmonizing the imperatives of environmental preservation and the varied demands of society, to forge a future in which water remains a sustainable and accessible resource for all.

3 Areas of Intelligent Water Management: Towards Innovative and Sustainable Use

3.1 Wastewater Recycling: Exploring the Benefits and Challenges

Wastewater recycling is emerging as an essential pillar of intelligent water management, offering an innovative solution to sustainability imperatives (Liu et al. 2021). By examining the benefits, this approach not only optimizes the use of water resources, but also reduces pressure on natural water sources. However, this section will not only outline the benefits, but also explore the challenges inherent in wastewater recycling, including health considerations, quality concerns, and social implications. This in-depth analysis aims to shed light on how smart water management can overcome these challenges to ensure effective and responsible wastewater recycling (Krishnan et al. 2022).

Wastewater recycling is emerging as an essential pillar of intelligent water management, offering an innovative solution to sustainability imperatives. By examining the benefits, this approach not only optimizes the use of water resources, but also reduces pressure on natural water sources. However, this approach goes beyond simply listing the benefits to explore the challenges inherent in wastewater recycling, including health considerations, quality concerns and social implications (Kamyab et al. 2023).

An exemplary case study in Singapore, where the NEWater program has been successfully implemented to recycle treated wastewater into a reliable source of drinking water. This initiative has demonstrated the technical feasibility and health safety of large-scale wastewater recycling, contributing to the island nation's water security (Banerjee et al. 2022; Fujii and Ray 2021).

In California, wastewater recycling projects, such as the Orange County Water District's Indirect Purification of Recycled Water program, have successfully alleviated pressure on freshwater sources while meeting the region's growing water needs. These initiatives have shown that wastewater recycling can be a pragmatic solution for building drought resilience and ensuring the sustainability of water supplies (Kenney 2019).

However, these case studies also highlight significant challenges. In Australia, the Toowoomba wastewater recycling project faced public resistance due to health concerns, underlining the crucial importance of communication and public education in promoting wastewater recycling (Mukherjee and Jensen 2020).

This in-depth analysis aims to highlight how smart water management, inspired by these case studies, can overcome these challenges. By developing advanced treatment technologies, setting rigorous sanitary standards and promoting public education, it becomes possible to ensure effective and responsible wastewater recycling, thus contributing to a sustainable and water-balanced future.

3.2 Water Distribution: Strategies for Intelligent Systems

Water distribution is emerging as one of the crucial areas of intelligent water management, highlighting innovative strategies for optimizing system efficiency. In this quest for innovative and sustainable use of water resources, strategies for intelligent distribution systems play a decisive role (Yanes et al. 2020).

Optimizing water distribution networks is one of the central strategies of intelligent management. Intelligent systems incorporate advanced sensors and real-time monitoring technologies to detect leaks and breaks, enabling rapid intervention and conservation of precious water resources. These reactive approaches not only help to conserve water, but also reduce the operational costs associated with repairs.

At the same time, smart water distribution networks enable dynamic, adaptive management of water flows. These systems use advanced algorithms to adjust pressure and distribution according to real-time needs, optimizing resource allocation and minimizing losses. This proactive approach is in line with sustainability, contributing to a more efficient use of water resources (Dogo 2019).

Advanced communication technologies, such as the Internet of Things (IoT), are also playing a major role in the transformation of water distribution systems. By enabling seamless connectivity between network components, they facilitate real-time data collection and informed decision-making. This enhanced connectivity promotes more agile and responsive management of distribution networks, strengthening resilience in the face of challenges such as outages and variations in demand (Dogo 2019).

Strategies for intelligent water distribution systems represent an essential area of intelligent water management. By integrating technology, real-time monitoring and dynamic network management, these approaches contribute to operational efficiency, the preservation of water resources and the long-term sustainability of distribution systems.

3.3 Rainwater Harvesting: Optimization Potential with AI, DL and IoT

Rainwater harvesting is emerging as a key area of intelligent water management, offering significant potential for optimization through the integration of artificial intelligence (AI), deep learning (DL) and the Internet of Things (IoT) (Fig. 2) (Krishnan et al. 2022). This innovative approach aims to harness natural resources efficiently while promoting sustainable water use (Darabi et al. 2021).

Integrating AI into rainwater harvesting systems enables proactive management based on predictive models. AI algorithms can analyze weather data, past rainfall patterns and water requirements to anticipate optimal harvesting periods. This accurate prediction enables recovery systems to be adjusted automatically, maximizing the efficiency of rainwater harvesting (Piro et al. 2023).

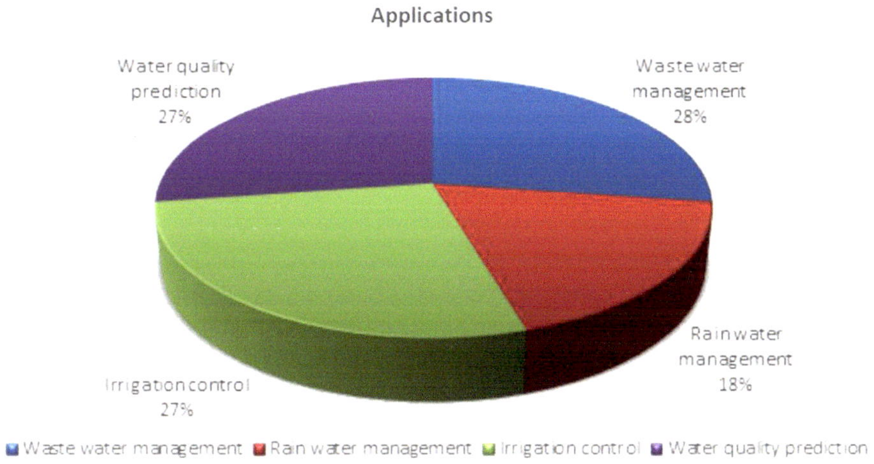

Fig. 2 Analyzed applications of water management techniques (Krishnan et al. 2022)

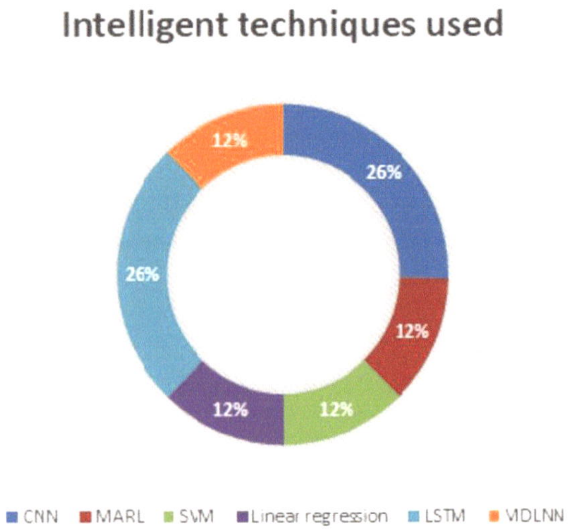

Fig. 3 Artificial intelligence techniques used in water management (Krishnan et al. 2022)

The use of DL offers another dimension of optimization, enabling systems to be constantly adapted to changing conditions. Deep neural networks can learn from seasonal variations, climatic changes and water consumption patterns, adjusting collection parameters to maximize yield under a variety of conditions.

IoT, by connecting rainwater harvesting devices to a smart network, enables real-time monitoring and remote management. IoT sensors can measure the quality of collected water, detect any anomalies, and signal maintenance requirements. This enhanced connectivity offers total visibility over recovery systems, enabling proactive management and rapid intervention in the event of problems (Xu et al. 2022).

By harnessing the combined potential of AI, DL and IoT, rainwater harvesting can be transformed into a highly efficient and adaptable process (Fig. 3) (Krishnan et al. 2022). These technologies offer the possibility of optimizing collection, minimizing losses and ensuring sustainable use of water resources, thus contributing to the achievement of sustainability goals in the field of water management.

3.4 Irrigation: Improving Farming Practices Through Advanced Technologies

Irrigation, as a key area of intelligent water management, paves the way for innovative and sustainable use of water resources, with significant potential for improving agricultural practices through advanced technologies. This convergence offers innovative solutions to the growing challenges of agriculture and water resource management (Gimpel et al. 2021).

The integration of advanced technologies such as smart sensors, automated irrigation systems and satellite remote sensing marks a transformative transition in irrigation. Intelligent sensors in the field collect real-time data on soil moisture, weather and other relevant parameters. This information is then used to automatically adjust irrigation systems, optimizing the amount of water applied according to actual crop needs (Martos et al. 2021).

Automated irrigation systems, based on advanced algorithms, enable precise, dynamic management of water supply. By continuously analyzing collected data, these systems can instantly adapt water flow, irrigation frequency and distribution to specific field conditions. This proactive approach considerably improves irrigation efficiency, minimizing wastage and ensuring optimal water use (Abioye et al. 2020).

Satellite remote sensing offers a broader perspective by enabling the monitoring of agricultural trends on a large scale. Satellite images can provide information on crop health, water requirements, and areas prone to water stress (Abioye et al. 2020). This overview enables farmers and water managers to make informed decisions to optimize irrigation practices on a regional scale.

Improving agricultural practices through advanced irrigation technologies is a fundamental pillar of smart water management. These innovations promote more efficient and sustainable use of water resources, while strengthening the resilience of agriculture in the face of challenges such as climate change and water scarcity.

4 Polyvalent Models and Algorithms: Towards Intelligent Adaptability in Water Management

4.1 Unique Data Requirements for Water Management

The exploration of versatile models and algorithms represents a crucial step towards intelligent adaptability in water management. However, to fully realize the potential of these approaches, it is essential to recognize and address the unique data requirements in the specific context of water management (Mahwayi 2016).

The data requirements for effective water management are multifaceted. Firstly, real-time data collection is crucial to enable agile response to rapid changes in water conditions. Sensors scattered across distribution networks, watersheds and water infrastructure provide essential information on water availability, quality and demand (Fitzgibbon et al. 2005).

Secondly, the spatial and temporal variability of data in water management requires models and algorithms capable of handling heterogeneous and evolving data sets. Water conditions can vary considerably according to season, geographical region and the specific characteristics of the water system. Models must be flexible enough to adapt to this diversity and produce relevant and accurate results (Aderemi et al. 2021).

Data quality is also of paramount importance. Reliable and accurate data sources are necessary to ensure the credibility of the models and algorithms used in water management. Errors or gaps in data can compromise the accuracy of forecasts and recommendations, underlining the importance of rigorous data quality management (Aderemi et al. 2021).

Finally, data security must be a major concern. With the increasing digitization of water management systems, the protection of sensitive data is becoming a priority. Robust security protocols must be put in place to prevent unauthorized access and ensure the integrity of information vital to water management.

Meeting the unique data requirements of water management is a crucial step towards exploiting the full potential of versatile models and algorithms. Efficient data acquisition, processing and security are fundamental to ensuring adaptive, intelligent and sustainable water management.

4.2 The Need for Versatile AI Models and Algorithms

The growing need for versatile artificial intelligence (AI) models and algorithms in water management has become imperative to achieve intelligent adaptability in this complex field. These versatile models are essential to meet the diversity of challenges linked to water availability, quality and distribution (Chau 2006; Tung and Yaseen 2020).

Firstly, the inherent variability of water systems requires models capable of adapting to changing conditions. Versatile AI algorithms can integrate real-time data from multiple sources, such as sensors, satellites and distribution networks, to dynamically adjust their predictions. This flexibility allows for seasonal fluctuations, sudden weather events and changes in water demand.

In addition, the complexity of interactions within aquatic ecosystems and distribution networks calls for versatile AI models capable of handling diverse data sets. These models can encompass biological, geographical and climatic factors, offering a holistic vision of water management. This versatile approach is crucial to understanding the complex and often non-linear relationships that govern hydrological cycles.

Versatile AI models also play a crucial role in optimizing water system operations. Whether forecasting water needs, detecting leaks in distribution networks, or managing flows in watersheds, these algorithms can adjust in real time to maximize operational efficiency.

Finally, in a context where water resources are increasingly under pressure, versatile AI models can contribute to strategic decision-making. By analyzing complex scenarios, these models can provide informed recommendations for proactive and sustainable management of water resources.

In conclusion, the need for versatile AI models and algorithms in water management is essential to enable intelligent adaptability to the complexity of contemporary water challenges. These versatile approaches offer a promising solution for sustainable and efficient management of water resources on a global scale.

4.3 Adaptability of Models to Varied Data Sets and Scenarios

In pursuing the goal of intelligent adaptability in water management, the adaptability of models to a variety of data sets and scenarios is proving to be a key element. This feature is essential to ensure that models and algorithms can effectively respond to the diversity of water conditions, needs and challenges encountered in different contexts.

The adaptability of models to varied data sets implies the ability to handle data from multiple sources, each characterized by its own dynamics and nature. Versatile models must be able to integrate data from sensors, satellites, distribution networks and other sources, while maintaining consistency and accuracy in their predictions. This flexibility is crucial to reflect the complexity of water systems across different geographical regions and climatic conditions.

Furthermore, the adaptability of models to a variety of scenarios requires the ability to anticipate and respond to unexpected changes in water conditions. Extreme weather events, seasonal variations and changes in human practices can all influence water availability and demand. Models need to be able to adjust their parameters in real time to provide accurate predictions and recommendations adapted to changing situations.

The use of machine learning algorithms, such as neural networks, enhances model adaptability by automatically identifying complex patterns in the data. These models can then evolve and adjust according to new examples, guaranteeing continuous responsiveness to emerging conditions.

In conclusion, the adaptability of models to a variety of data sets and scenarios is an essential component in the quest for intelligent, adaptive water management. By developing versatile models capable of dynamically adjusting to changing conditions, it becomes possible to provide more robust and effective solutions for managing water resources in diverse contexts.

5 Integrating AI, DL and IoT in the Design of an Intelligent Water Management System

The synergistic integration of Artificial Intelligence (AI), Deep Learning (DL), and Internet of Things (IoT) technologies offers a novel perspective for the design of intelligent water management systems. This section will explore how these techniques can work together holistically to address the complex challenges of water management.

5.1 Synergy of AI, DL and IoT Techniques

The integration of artificial intelligence (AI), deep learning (DL) and the Internet of Things (IoT) in the design of an intelligent water management system marks a significant advance towards a holistic and innovative approach. The synergy of these techniques offers promising prospects for tackling the complex challenges of water resource management in an efficient and adaptive way (Ghazal et al. 2021).

The synergy between AI, DL and IoT relies on the ability to exploit the complementary strengths of each field. AI, as a global concept, encompasses a variety of techniques, including machine learning and DL. These methods can analyze huge datasets, detect complex patterns and make autonomous decisions, thus contributing to intelligent decision-making in water management.

DL, in particular, plays a crucial role in enabling systems to understand and learn from data autonomously. Deep neural networks can extract complex features from water data, facilitating trend forecasting, anomaly detection and process optimization.

IoT enriches this synergy by providing connectivity between physical objects and IT systems. Sensors and IoT devices scattered across distribution networks, watersheds and water infrastructure collect data in real time. This data then feeds AI and DL models, enabling proactive and reactive management of water resources.

The combination of these technologies offers intelligent, adaptive water management on many levels. For example, IoT sensors can monitor water quality in real time, detecting potentially harmful variations. AI models can analyze this data to anticipate future water needs and optimize distribution. DL can be used to improve the accuracy of forecasts based on complex models. The synergy between AI, DL and IoT represents a powerful integrated approach to the design of intelligent water management systems. This combination offers an enhanced ability to understand, anticipate and react to changing water conditions, thus contributing to a more efficient and sustainable use of water resources.

5.2 Examples of Successful Implementation and Impact on Sustainability

The successful integration of artificial intelligence (AI), deep learning (DL) and the Internet of Things (IoT) in the design of intelligent water management systems has already demonstrated significant impacts on sustainable development. Several examples of successful implementation illustrate how these converging technologies can revolutionize water resource management.

Singapore has successfully implemented the NEWater program, using AI to optimize the wastewater purification process. IoT sensors monitor water quality in real time, while DL is employed to improve the efficiency of the treatment process. This approach has enabled Singapore to maximize the use of local water resources and enhance the sustainability of its water supply (Banerjee et al. 2022; Fujii and Ray 2021).

In California, intelligent water management systems use AI and IoT to monitor water levels, anticipate seasonal demands and adjust irrigation flows. These technologies have helped optimize water use in agriculture, minimize wastage and address the challenges posed by recurring droughts, supporting the sustainability of water resources (Kenney 2019).

Barcelona has implemented a smart water distribution network using IoT sensors to detect leaks and monitor pressure in real time. AI analyzes this data to forecast water requirements and dynamically adjust distribution. This approach has not only reduced water losses, but also contributed to more efficient and sustainable water management in the city (Adedeji et al. 2022).

Rwanda has adopted AI-based predictive systems to anticipate rainfall patterns. These models use IoT and satellite data to improve the accuracy of weather forecasts. This information is then used to guide agricultural decisions, improving water management in the farming sector and strengthening resilience in the face of climate change (Pierre et al. 2023).

The impact of these examples goes beyond operational efficiency, also contributing to sustainable development goals by promoting responsible and equitable use of water resources, reducing environmental impacts and strengthening resilience to climate change. These success stories underline the crucial importance of the intelligent integration of AI, DL and IoT for sustainable water management on a global scale.

5.3 Potential Benefits, Challenges and Future Prospects

The integration of artificial intelligence (AI), deep learning (DL) and the Internet of Things (IoT) in intelligent water management offers considerable potential benefits. These converging technologies enable optimization of water resources, reduce losses through real-time leak detection, contribute to environmental sustainability by preserving water quality, strengthen resilience in the face of climate change, and improve the operational efficiency of water infrastructures.

However, this integration is not without its challenges. The complexity of the data, the need for significant initial investment, data security risks and the challenges of social acceptance represent major obstacles. The diversity and volume of data require advanced processing capabilities, while data security becomes crucial with the increased connectivity of systems.

Future prospects for the integration of AI, DL and IoT in water management are promising. The evolution towards autonomous systems capable of making complex decisions in real time is envisaged. In addition, integration with other emerging technologies, such as blockchain, can enhance traceability and data security. International collaboration and the development of common standards can encourage the sharing of best practices.

Despite today's challenges, the intelligent integration of AI, DL and IoT offers innovative solutions for sustainable water management. Continued progress in the field promises even more intelligent, autonomous and adaptive systems, contributing to the efficient preservation of water resources on a global scale.

6 Case Studies and Statistical Analyses in the Application of AI, DL and IoT in Water Management

To illustrate concretely the effectiveness of Artificial Intelligence (AI), Deep Learning (DL), and the Internet of Things (IoT) in water management, several case studies reveal the substantial impact of these technologies on various aspects of the water cycle. These concrete examples, backed by rigorous statistical analysis, reinforce the perspective that AI, DL, and IoT are not just theoretical concepts, but pragmatic tools that are tangibly transforming contemporary water management. These emerging technologies are demonstrating their potential to solve major challenges while contributing to the efficiency, sustainability and preservation of water resources.

The application of artificial intelligence (AI), deep learning (DL) and the Internet of Things (IoT) in water management is supported by several case studies illustrating the concrete benefits of these technologies. In one metropolis, IoT sensors are deployed along the distribution network, collecting real-time data on water pressure, flow and quality. AI analyzes this data to anticipate needs, optimize distribution, and quickly detect leaks, thus contributing to proactive and efficient network management.

Another example is in an agricultural region, where IoT devices collect weather data for AI models based on deep learning. These models forecast rainfall, enabling proactive adaptation of irrigation practices for efficient water use in agriculture and minimization of environmental impacts. In a flood risk management context, IoT sensors monitor water levels along a watershed. AI analyzes this data to anticipate risks, triggering real-time alerts and enabling the implementation of preventive measures, such as the controlled release of water from dams.

Predicting real-time water treatment parameters is challenging. A data-driven approach using anaerobic membrane bioreactors (AnMBRs) for municipal wastewater treatment. Deep learning structures are employed, achieving a 97.44% accuracy and rapid calculations (<1 s). This enhances AnMBR treatment prediction, reducing pollutants and costs. Wastewater treatment facilities (WWTP) generate sludge and greenhouse gases. Optimization using multi-agent deep reinforcement learning minimizes environmental impacts, reducing cost, energy, and greenhouse gas consumption. It emphasizes the importance of considering effect factors during WWTP retrofitting based on resources.

Deep learning models excel in water quality prediction. Focuses on Bi-LSTM for time-series water quality data in Yamuna River. The model achieves superior accuracy (MSE, RMSE, MAE, MAPE) compared to other models, showcasing the effectiveness of smart sensors and auto deep learning (Auto DL). Auto DL provides flexibility and efficiency in real-time applications. Water quality variations due to floods, rains, and drainage affect data quality. Auto DL outperforms conventional models, offering simplicity and reduced time complexity in practical applications.

Rainfall prediction is crucial for water harvesting. ML and DL models were used for accuracy assessment. DL outperforms ML, indicating its suitability for irregular datasets like rainfall. Automated rainwater harvesting systems are proposed, combining computer vision and deep learning for optimal tank placement. Time-series analysis is crucial due to climatic variations affecting data quality.

Groundwater overuse for agriculture poses global threats. Deep learning models classify soil texture and irrigation control systems, providing efficient water management. Evaporation prediction enhances irrigation control, with hybrid LSTM models achieving accurate estimations. Challenges include data quality issues from sensor failures and cyber threats.

These generic examples illustrate how these converging technologies can be successfully applied in a variety of contexts to improve water resource management, demonstrating their growing relevance in the field of water infrastructure sustainability and efficiency. For specific case studies and in-depth analyses, it is recommended to refer to specialized sources in the field of water management.

7 Conclusion: Towards Innovative and Sustainable Water Management

In conclusion, the exploration of water management through the application of Artificial Intelligence (AI), Deep Learning (DL), and the Internet of Things (IoT) reveals significant findings and enriching perspectives.

Synergies between AI, DL and IoT enable proactive and adaptive water management. The case studies analyzed demonstrate that these technologies can be fruitfully integrated in a variety of contexts, from real-time monitoring of distribution networks to anticipating climatic variations affecting water quality. The results of statistical analyses support increased operational efficiency, with significant reductions in water losses, associated costs, and environmental risks.

The potential for developing a successful and sustainable water management system is clearly evident. The judicious integration of these intelligent technologies paves the way for systems capable of dynamically adjusting their operations, optimizing water distribution, and preserving the quality of this vital resource. This technological convergence not only offers solutions to today's water management challenges, but also serves as a solid foundation for a more resilient and efficient future approach.

Ultimately, this exploration underscores the crucial importance of technological innovation in water management to meet our planet's growing needs. The intelligent integration of AI, DL, and IoT paves the way for modern, efficient, and sustainable water management, essential to ensure the long-term preservation of this vital resource.

Competing Interests The author declares that there is no conflict of interests regarding the publication of this manuscript. In addition, the ethical issues, including plagiarism, informed consent, misconduct, data fabrication and/or falsification, double publication and/or submission, and redundancy has been completely observed by the authors.

References

Abioye EA et al (2020) A review on monitoring and advanced control strategies for precision irrigation. Comput Electron Agric 173:105441

Achoukhi I et al (2023) Investigating microplastics in the Mediterranean coastal areas-case study of Al-Hoceima Bay, Morocco. J Ecol Eng 24(5):12–31

Adedeji KB et al (2022) Towards digitalization of water supply systems for sustainable smart city development—water 4.0. Appl Sci 12(18):9174

Aderemi BA et al (2021) A review of groundwater management models with a focus on IoT-based systems. Sustainability 14(1):148

Banerjee C, Bhaduri A, Saraswat C (2022) Digitalization in urban water governance: case study of Bengaluru and Singapore. Front Environ Sci 10:266

Benaissa C et al (2020) Hydro-chemical and bacteriological study of some sources of groundwater in the GHIS-NEKOR and the BOKOYA aquifers (AL HOCEIMA, MOROCCO). In: Proceedings of the 4th edition of international conference on Geo-IT and water resources

Benaissa C et al (2022) Assessment of Water quality using water quality index – case study of Bakoya Aquifer, Al Hoceima, Northern Morocco. Ecol Eng & Environ Technol 23(4):31–44

Berthet A, Vincent A, Fleury P (2021) Water quality issues and agriculture: an international review of innovative policy schemes. Land Use Policy 109:105654

Bhateria R, Jain D (2016) Water quality assessment of lake water: a review. Sustain Water Resour Manag 2:161–173

Bogardi JJ et al (2012) Water security for a planet under pressure: interconnected challenges of a changing world call for sustainable solutions. Curr Opin Environ Sustain 4(1):35–43

Bouhout S et al (2023) Appraisal of groundwater quality status in the Ghiss-Nekor coastal plain. J Ecol Eng 24(10)

Bourjila A et al (2021) Groundwater potential zones mapping by applying GIS, remote sensing and multi-criteria decision analysis in the Ghiss basin, northern Morocco. Groundw Sustain Dev 15:100693

Bourjila A et al (2023) Mapping the spatiotemporal evolution of seawater intrusion in the Moroccan coastal aquifer of Ghiss-Nekor using GIS-based modeling. Water Cycle 4:104–119

Brown CM et al (2015) The future of water resources systems analysis: toward a scientific framework for sustainable water management. Water Resour Res 51(8):6110–6124

Chau K-W (2006) A review on integration of artificial intelligence into water quality modelling. Mar Pollut Bull 52(7):726–733

Darabi H et al (2021) Efficient rainwater harvesting planning using socio-environmental variables and data-driven geospatial techniques. J Clean Prod 311:127706

De Fraiture C, Molden D, Wichelns D (2010) Investing in water for food, ecosystems, and livelihoods: an overview of the comprehensive assessment of water management in agriculture. Agric Water Manag 97(4):495–501

Dimane F, El Hammoudani Y (2021) Assessment of quality and potential reuse of wastewater treated with conventional activated sludge. Mater Today: Proc 45:7742–7746

Dogo EM et al (2019) Blockchain and internet of things-based technologies for intelligent water management system. In: Artificial intelligence in IoT, pp 129–150

Eggimann S et al (2017) The potential of knowing more: a review of data-driven urban water management. Environ Sci Technol 51(5):2538–2553

El Abdouni A et al (2021) Physicochemical characterization of wastewater from the Al-Hoceima slaughterhouse in Morocco. Casp J Environ Sci 19(3):423–429

El Hammoudani Y, Dimane F (2020) Assessing behavior and fate of micropollutants during wastewater treatment: statistical analysis. Environ Eng Res

El Hammoudani Y, Dimane F (2021) Occurrence and fate of micropollutants during sludge treatment: case of Al-Hoceima WWTP, Morocco. Environ Chall 5(1):1–8

El Hammoudani Y, Dimane F, El Ouarghi H (2019) Fate of selected heavy metals in a biological wastewater treatment system. In: Euro-Mediterranean conference for environmental integration. Springer

El Hammoudani Y, Dimane F, El Ouarghi H (2021) Removal efficiency of heavy metals by a biological wastewater treatment plant and their potential risks to human health. Environ Eng Manag J 20(6):995–1002

El Hammoudani Y et al (2023) Assessment of groundwater quality in the lower Wadi of the Nekor valley, Al-Hoceima-Morocco. Environ Eng & Manag J (EEMJ) 22(3)

Elabdouni A et al (2020) Olive mill wastewater (OMW) production in the province of Al-Hoceima (Morocco) and their physico-chemical characterization by mill types. Mater Today: Proc 27:3145–3150

Elabdouni A et al (2022) Removal of organic matter and polyphenols in the olive oil mill wastewater by coagulation-flocculation using aluminum sulfate and lime. Moroccan J Chem 10(1):191–202

Fitzgibbon K et al (2005) Innovative water monitoring

Fujii T, Ray R (2021) Singapore as a sustainable city: past, present, and the future. The Singapore Economy. Routledge, pp 143–192

Ghazal TM et al (2021) IoT for smart cities: machine learning approaches in smart healthcare—a review. Future Internet 13(8):218

Gimpel H et al (2021) Designing smart and sustainable irrigation: a case study. J Clean Prod 315:128048

Hassan MM et al (2021) Efficient prediction of water quality index (WQI) using machine learning algorithms. Hum-Cent Intell Syst 1(3–4):86–97

Irfeey AMM et al (2023) Water valuation in urban settings for sustainable water management. Water 15(17):3105

Javed AR et al (2022) Future smart cities: requirements, emerging technologies, applications, challenges, and future aspects. Cities 129:103794

Kabote SJ, John P (2017) Water governance in Tanzania: performance of governance structures and institutions

Kamyab H et al (2023) The latest innovative avenues for the utilization of artificial Intelligence and big data analytics in water resource management. Results Eng 101566

Kenney S (2019) Purifying water: Responding to public opposition to the implementation of direct potable reuse in California. UCLA J Environ & Policy 37:85

Krishnan SR et al (2022) Smart water resource management using artificial intelligence—a review. Sustainability 14(20):13384

Liu Q, Yang L, Yang M (2021) Digitalisation for water sustainability: barriers to implementing circular economy in smart water management. Sustainability 13(21):11868

Lowe M, Qin R, Mao X (2022) A review on machine learning, artificial intelligence, and smart technology in water treatment and monitoring. Water 14(9):1384

Mahwayi MP (2016) Modelling groundwater monitoring for rural areas using cloud computing. University of Johannesburg (South Africa)

Martos V et al (2021) Ensuring agricultural sustainability through remote sensing in the era of agriculture 5.0. Appl Sci 11(13):5911

Mondejar ME et al (2021) Digitalization to achieve sustainable development goals: steps towards a smart green planet. Sci Total Environ 794:148539

Mukherjee M, Jensen O (2020) Making water reuse safe: a comparative analysis of the development of regulation and technology uptake in the US and Australia. Saf Sci 121:5–14

Mustapha UF et al (2021) Sustainable aquaculture development: a review on the roles of cloud computing, internet of things and artificial intelligence (CIA). Rev Aquac 13(4):2076–2091

Newson M (2008) Land, water and development: sustainable and adaptive management of rivers. Routledge

Pappalardo C (2018) Innovation through traditional water knowledge: an approach to the water crisis. Geo Environ Rev 31:627

Pierre N et al (2023) AI based real-time weather condition prediction with optimized agricultural resources. Eur J Technol 7(2):36–49

Piro P et al (2019) Smart and innovative systems for urban flooding risk management. In: 2023 international conference on Information and Communication Technologies for Disaster Management (ICT-DM). IEEE

Rijsberman FR (2006) Water scarcity: fact or fiction? Agric Water Manag 80(1–3):5–22

Sadoff C, Muller M (2009) Water management, water security and climate change adaptation: early impacts and essential responses. Global Water Partnership Stockholm

Schlueter M et al (2012) New horizons for managing the environment: a review of coupled social-ecological systems modeling. Nat Resour Model 25(1):219–272

Scott CA et al (2021) Water security and the pursuit of food, energy, and earth systems resilience. Putting water security to work. Routledge, pp 97–116

Sharan A, Datta B, Lal A (2023) Integrating numerical modelling and scenario-based sensitivity analysis for saltwater intrusion management: case study of a complex heterogeneous island aquifer system. Environ Monit Assess 195(5):553

Subramani A, Jacangelo JG (2015) Emerging desalination technologies for water treatment: a critical review. Water Res 75:164–187

Sun AY, Scanlon BR (2019) How can big data and machine learning benefit environment and water management: a survey of methods, applications, and future directions. Environ Res Lett 14(7):073001

Taher M, Mourabit T, Bourjila A, Saadi O, Errahmouni A, El Marzkioui F, El Mousaoui M (2022) An estimation of soil erosion rate hot spots by integrated USLE and GIS methods: a case study of the Ghiss Dam and Basin in Northeastern Morocco. Geomat Environ Eng 2(16):95–110

Tung TM, Yaseen ZM (2020) A survey on river water quality modelling using artificial intelligence models: 2000–2020. J Hydrol 585:124670

Wang X et al (2023) Deep-learning-based water quality monitoring and early warning methods: a case study of ammonia nitrogen prediction in rivers. Electronics 12(22):4645

Xu WD et al (2022) Real-time controlled rainwater harvesting systems can improve the performance of stormwater networks. J Hydrol 614:128503

Yanes AR, Martinez P, Ahmad R (2020) Towards automated aquaponics: a review on monitoring, IoT, and smart systems. J Clean Prod 263:121571

Zevenbergen C, Fu D, Pathirana A (2018) Transitioning to sponge cities: challenges and opportunities to address urban water problems in China. Water 10(9):1230

Zhou X et al (2019) Deep learning identifies accurate burst locations in water distribution networks. Water Res 166:115058